Electronic Alarm Circuits Manual

Other Books by R. M. Marston

Audio IC Circuits Manual
CMOS Circuits Manual
110 Semiconductor Projects for the Home Constructor
110 Integrated Circuit Projects for the Home Constructor
110 Thyristor Projects using SCRs and Triacs
110 Waveform Generator Projects for the Home Constructor
Op-Amp Circuits Manual
Optoelectronics Circuits Manual
Power Control Circuits Manual
Timer/Generator Circuits Manual

Electronic Alarm Circuits Manual

R. M. Marston

Heinemann Newnes

Heinemann Newnes
An imprint of Heinemann Professional Publishing Ltd
Halley Court, Jordan Hill, Oxford OX2 8EJ

OXFORD LONDON MELBOURNE AUCKLAND SINGAPORE
IBADAN NAIROBI GABORONE KINGSTON

First published 1988
Reprinted 1990

British Library Cataloguing in Publication Data

Marston, R. M. (Raymond Michael), 1937–
Electronic alarm circuits manual.
1. Electronic equipment. Alarm circuits.
Projects
I. Title
621.389′2

ISBN 0 434 91214 X

Cover circuit supplied by
RS Components

Printed and bound in Great Britain by
Biddles Ltd, Guildford and King's Lynn

CONTENTS

PREFACE

Electronic alarms have many applications in the home, in industry, and in the car. They can be designed to be activated by physical contact or by body proximity, or by variations in light or heat levels, or in voltage, current, resistance or some other electrical property. They can be designed to give high-level audible outputs, as in the case of burglar alarms, or low-level visual outputs, as in the case of instrumentation alarms. One hundred and forty useful alarm circuits, of a variety of types, are shown in this volume. The operating principle of each one is explained in concise but comprehensive terms, and brief construction notes are given where necessary.

The volume is split into eight chapters. The first five are devoted to alarm circuits that can be used in the home and in industry, including contact-operated alarms, burglar alarms, temperature alarms, light-sensitive alarms, proximity and touch alarms, power-failure alarms, and sound and vibration alarms. Chapter 6 is devoted to automobile alarm circuits, and gives details of immobilisers and anti-theft alarms, ice-hazard alarms, overheat alarms, and low-fuel-level alarms. Chapter 7 is devoted to instrumentation alarm circuits, and shows alarms that can be activated by a.c. or d.c. currents or voltages or by resistance. The final chapter deals with IR (infra-red) light-beam alarms, and presents a selection of outstandingly useful circuits.

All the circuits described in this volume have been designed, built and fully evaluated by the author, and will be of equal interest to the electronics amateur, student and engineer. They are designed around a variety of types of readily available semi-conductor devices; most are designed around standard bipolar transistors, or an 8-pin type 741 operational amplifier, or a type 4001B (or CD4001) quad 2-input CMOS (COS/MOS) NOR gate digital IC. Some circuits have an SCR output stage, and are used to activate an alarm bell or buzzer directly; in this latter case the SCR is a type C106Y1 or equivalent, and is meant to be used in conjunction with a self-interrupting bell or buzzer with a mean current rating of less than 2 A and a voltage rating that is 1.5 V less than the circuit's power supply.

The outlines and pin connections of all semi-conductors mentioned in the volume are given in the appendix, as an aid to construction. Unless otherwise stated, all resistors used in the circuits are standard half-watt types.

CHAPTER 1

CONTACT-OPERATED ALARM CIRCUITS

Contact-operated alarms can be simply described as alarm systems that are activated by the opening or closing of a set of electrical contacts. These contacts may take the form of a simple push-button switch, a pressure-pad switch, or a reed-relay, etc. The actual alarms may be designed to give an audible loudspeaker output, an alarm-bell output, or a relay output that can be used to operate any kind of audible or visual warning device. The alarm system can be designed to give non-latch, self-latch, or one-shot operation.

Contact-operated alarms have many practical applications in the home and in industry. They can be used to attract attention when someone operates a push switch, to give an automatic warning when someone opens a door or window or treads on a pressure-pad, or to give an alarm indication when a piece of machinery moves beyond a preset limit and activates a microswitch, etc. A wide range of practical contact-operated alarm circuits are described in this chapter.

Alarm-bell & relay-output alarm circuits

The simplest possible type of contact-operated alarm circuit consists of an alarm bell in series with a normally-open (n.o.) switch, the combination being wired across a suitable battery supply, as shown in *Figure 1.1.* Any number of n.o. switches can be wired in parallel, so that the alarm operates when any one or more of these switches is closed. This type of circuit gives an inherently non-latch form of operation, and has the great advantage of drawing zero standby current from its supply battery.

A major disadvantage of the simple *Figure 1.1* circuit is that it passes the full alarm bell current through the n.o. operating switches, so these

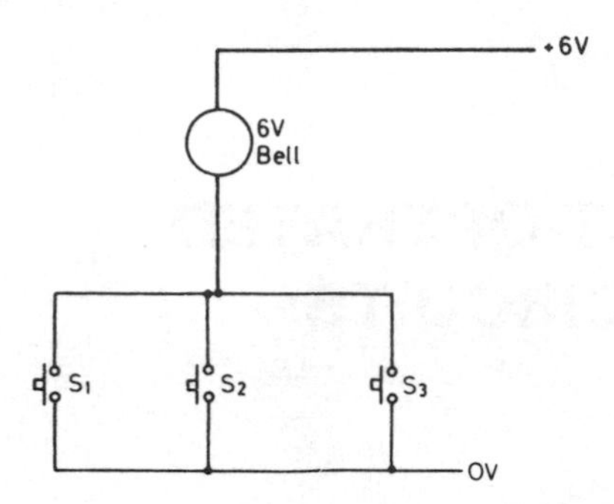

Figure 1.1 Simple non-latch close-to-operate alarm circuit

switches must be fairly robust types. One simple way round this problem is shown in *Figure 1.2a.*

Here, the n.o. operating switches are wired in series with the coil of a 6 V relay, and the relay contacts are wired in series with the alarm bell; both combinations are wired across the same 6 V supply. Thus when the switches are open the relay is off, so the relay contacts are open and the alarm bell is inoperative. When one or more of the switches is

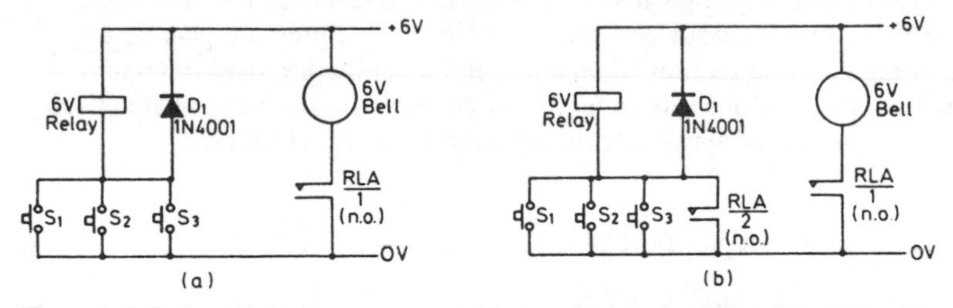

Figure 1.2 Relay-aided close-to-operate alarms: (a) non-latching; (b) self-latching

closed the relay turns on, so its contacts close and operate the alarm bell. Note in the latter case that the switches pass a current equal to that of the relay coil only, and can thus be fairly delicate types such as reed-relays, etc. Also note that a silicon diode is wired across the relay coil, to protect the switches against damage from the back e.m.f. of the coil.

The *Figure 1.2a* circuit gives a non-latch form of operation, in which the alarm operates only while one or more of the operating switches is closed. If desired the circuit can be made self-latching, so that the relay and the alarm lock on as soon as one or more of the n.o. switches is closed, by wiring a spare set of n.o. relay contacts in parallel with the n.o. operating switches, as in *Figure 1.2b*.

An alternative solution to the switch-current problem is shown in *Figure 1.3a*. Here, a general-purpose silicon controlled rectifier (SCR)

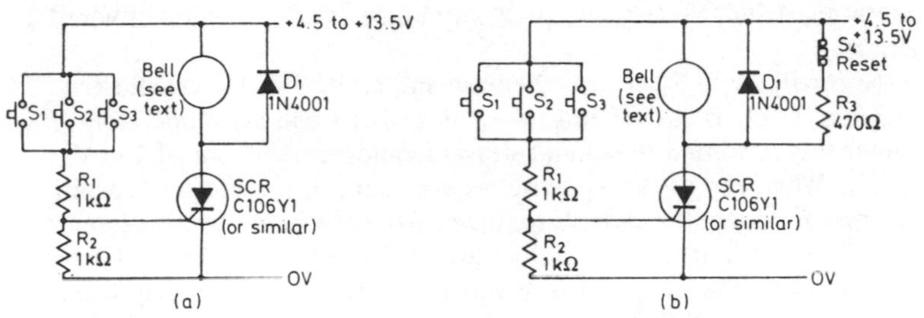

Figure 1.3 SCR-aided close-to-operate alarms: (a) non-latching; (b) self-latching

is wired in series with an inexpensive self-interrupting alarm bell, and the SCR has its gate current derived from the positive supply line via current-limiting resistor R_1 and via the n.o. operating switches. R_1 and the switches pass a current of only a few milliamps.

Normally, when the switches are open, the SCR is off and the alarm bell is inoperative. When one or more of the switches is closed, gate current is applied to the SCR via R_1, so the SCR turns on and the alarm bell operates. Since the operating current of a self-interrupting alarm bell is applied in a series of pulses via the built-in self-energising vibration contacts of the bell, the simple *Figure 1.3a* SCR circuit gives a non-latch type of operation when it is used with this type of alarm device.

If desired, the *Figure 1.3a* circuit can be made to give self-latch operation by simply wiring a shunt resistor across the bell, as shown in *Figure 1.3b*, so that the SCR current does not fall below its latching value when the bell goes into the self-interrupting mode.

Note that the SCR used in the *Figure 1.3* circuits has a current rating of only 2 A, so the alarm bell must be selected with this point in mind. Alternatively, SCRs with higher current ratings can be used in place of the device shown, but this modification may also necessitate changes in the R_1 and R_3 values of the circuit.

A major weakness of the *Figure 1.1* to *1.3* circuits is that they do not give a 'fail-safe' form of operation, and give no indication of a fault condition if a break occurs in the contact-switch wiring. This snag is overcome in circuits that are designed to be activated via normally-closed (n.c.) switches, and a typical circuit of this type is shown in *Figure 1.4.*

Here, the coil of a 12 V relay is wired in series with the collector of transistor Q_1, and bias resistor R_1 is wired between the positive supply line and the base of the transistor. The alarm bell is wired across the supply lines via a set of n.o. relay contacts, and the n.c. operating switch is wired between the base and emitter of the transistor. Note that operating switch S_1 may comprise any number of n.c. switches wired in series.

Normally, with S_1 closed, the base and emitter of Q_1 are shorted together, so Q_1 is cut off and the relay and the bell are inoperative. Under this condition the circuit draws a quiescent current of 1 mA via R_1. When any of the S_1 switches are open, or if a break occurs in the switch wiring, the short is removed from the base–emitter junction of Q_1, and current flows into the base of the transistor via R_1. Under this condition the transistor is driven to saturation, so the relay turns on and the alarm bell operates as the relay contacts close. The basic *Figure 1.4* circuit can, if desired, be made to give self-latch operation by wiring a spare set of n.o. relay contacts between the collector and emitter of Q_1, as shown dotted in the diagram.

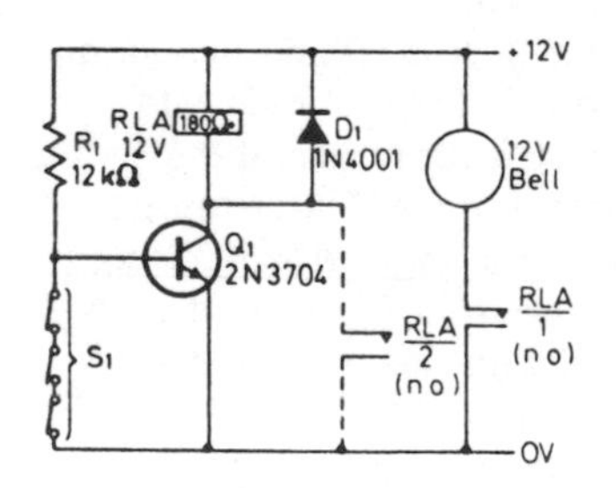

Figure 1.4 Simple break-to-operate alarm draws 1 mA standby current

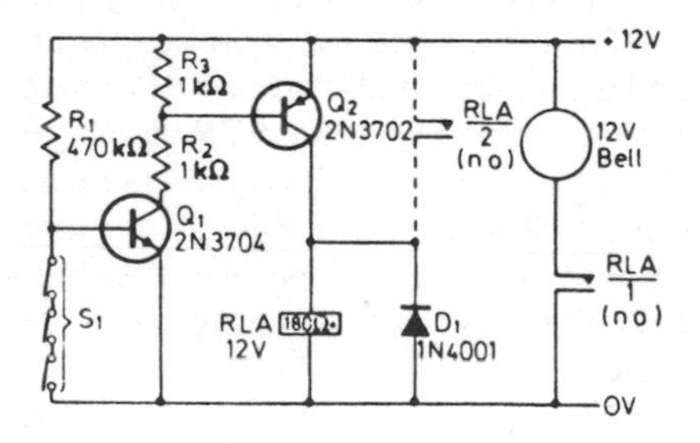

Figure 1.5 Improved break-to-operate alarm draws 25 μA standby current

Thus the *Figure 1.4* circuit gives automatic fail-safe operation, but draws the fairly high quiescent or standby current of 1 mA. This standby current can readily be reduced to a mere 25 μA by interposing a two-transistor amplifier stage between the contact switch and the relay, as shown in *Figure 1.5.*

Here, the base current of Q_2 is derived from the collector of Q_1 via R_2, and the base current of Q_1 is derived from the positive supply line via R_1. Consequently, when S_1 is closed both Q_1 and Q_2 are cut off, so the relay and the alarm bell are inoperative, but when S_1 is open both transistors are driven to saturation and the relay and the alarm bell are driven on. The circuit can be made self-latching by wiring a spare set of n.o. relay contacts between the collector and emitter of Q_2, as shown dotted in the diagram.

If desired, the standby current of the *Figure 1.5* circuit can be reduced to a mere 1 μA or so by using an inverter-connected COS/MOS gate in place of Q_1, as shown in *Figure 1.6.* The gate used here is taken

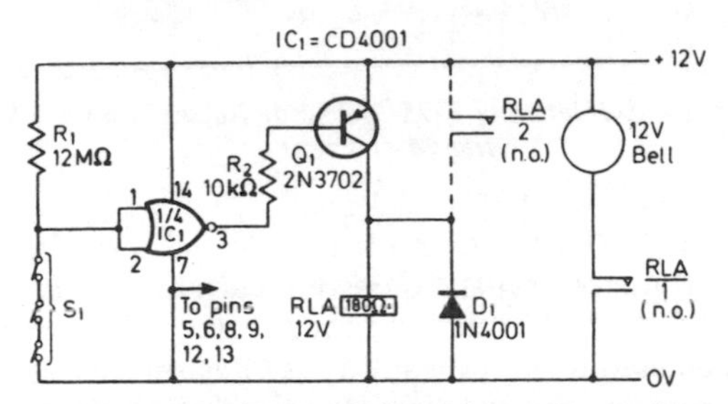

Figure 1.6 COS/MOS-aided break-to-operate alarm draws 1 μA standby current

from a CD4001 quad two-input NOR gate IC, and the remaining three unused gates of the device are disabled by taking their input pins to ground, as indicated. The inverter-connected COS/MOS gate has a virtually infinite input impedance, so R_1 can be given almost any resistance value. Ultimately, the minimum standby current of the circuit is limited only by the value of R_1 and the leakage current of Q_1.

The *Figure 1.6* circuit can be made self-latching by wiring a spare set of n.o. relay contacts between the collector and emitter of Q_1, as shown dotted in the diagram. Alternatively, the circuit can be made to self-latch by connecting two of the NOR gates of the CD4001 IC as a bistable multivibrator, as shown in *Figure 1.7.* Note that the remaining two unused gates of the IC are disabled by taking their inputs to ground.

The operation of the *Figure 1.7* circuit is such that the output of the COS/MOS bistable circuit goes low and self-latches if S_1 is momentarily opened or if the S_1 contact leads are interrupted. As the output of the bistable goes low it turns on the relay and the alarm bell via Q_1. Once the bistable has latched the alarm bell into the 'on' state, it can be reset into the standby 'off' mode by closing S_1 and momentarily operating 'reset' switch S_2, at which point the output

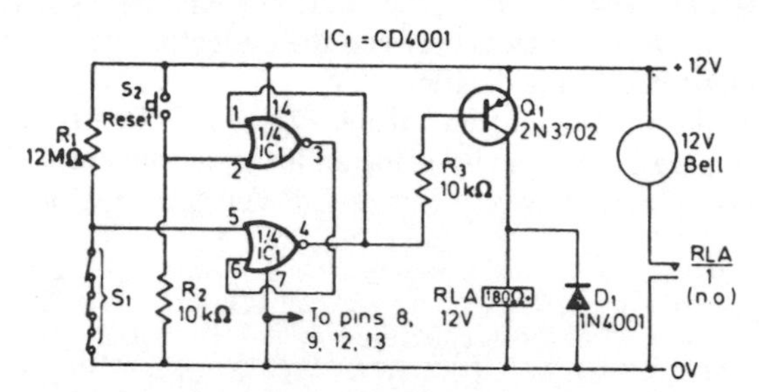

Figure 1.7 Self-latching COS/MOS-aided alarm draws 1 μA standby current

of the bistable latches back into the high state and turns Q_1 off. The *Figure 1.7* circuit draws a typical quiescent current of only 1 μA or so.

Note that, if desired, the relay-output circuits of *Figures 1.2* and *1.4* to *1.7* can be used to activate any type of alarm devices via their n.o. relay contacts, and are not restricted to use with alarm bells only.

Loudspeaker-output alarm circuits

Contact-operated alarm circuits can be designed to produce electronically-generated alarm signals directly into loudspeakers. Such systems can be made to produce a variety of sounds, at a variety of power levels, and may be designed around a number of types of semiconductor devices.

The most useful type of semiconductor device for this particular application is the COS/MOS digital integrated circuit. In particular, the CD4001 quad two-input NOR gate IC has the outstanding advantages of drawing virtually zero standby current, of having a virtually infinite input impedance, of tolerating a wide range of supply-rail voltages, and of being so versatile that it can be used in a whole range of waveform-generating applications. *Figures 1.8* to *1.16* show a variety of ways of using a single CD4001 IC to make contact-operated loudspeaker-output alarm circuits.

Figure 1.8 shows the circuit of a low-power contact-operated 800 Hz (monotone) alarm generator. Here, two of the gates of a CD4001 are connected as a gated 800 Hz astable multivibrator, and the remaining two unused gates of the device are disabled by taking their inputs to ground. The action of this astable is such that it is inoperative, with its pin 4 output terminal locked to the positive supply rail, when its pin 1 input terminal is high (at positive-rail voltage), but is operative and acting as a square-wave generator when its input is low (tied to the zero-volts line). The frequency of the astable is determined by the values of R_1 and C_1.

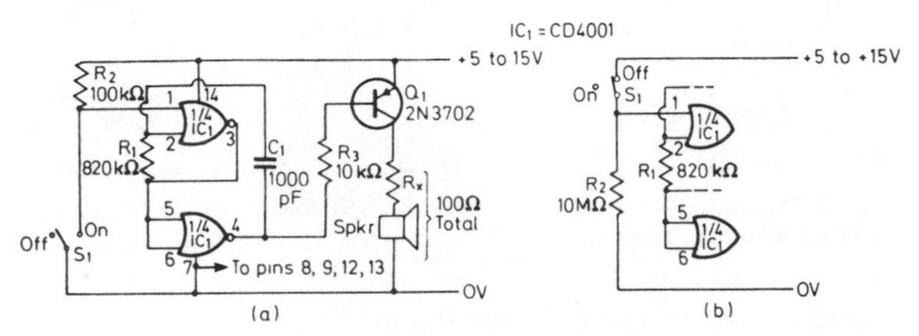

Figure 1.8 (a) Low-power 800 Hz close-to-operate alarm generator. (b) Modification for break-to-operate version

Thus, when the input terminal is high, zero base-current is fed to Q_1, so the circuit is inoperative and passes only a small leakage current; but when the input is low the astable is operative and generates a square-wave 'tone' signal in the speaker via Q_1. Note therefore that the circuit can be activated via n.o. contacts by using the input connections shown in *Figure 1.8a,* or by n.c. contacts by using the input connections shown in *Figure 1.8b.* In the latter case the circuit draws a standby current of roughly 1 μA via bias resistor R_2.

The basic *Figure 1.8* circuit is intended for low-power applications only, and can be used with any speaker in the range 3–100 Ω, and with any supply in the range 5–15 V. Note that resistor R_x must be wired in series with the speaker, and must be chosen so that the total series resistance of R_x and the speaker approximates 100 Ω, to keep the dissipation of Q_1 within acceptable limits. The actual power-output level of the circuit depends on the individual values of speaker impedance and supply voltage that are used, but is of the order of only a few milliwatts.

If desired, the power output can be boosted to a more useful level by using the medium-power output-stage circuit of *Figure 1.9.* The output power of this circuit again depends on the supply-rail and speaker-impedance values used, and may vary from 0.25 W when a

25 Ω speaker is used with a 5 V supply, to 11.25 W when a 5 Ω speaker is used with a 15 V supply. Alternatively, the output level can be boosted to about 18 W by using the high-power output stage of *Figure 1.10.*

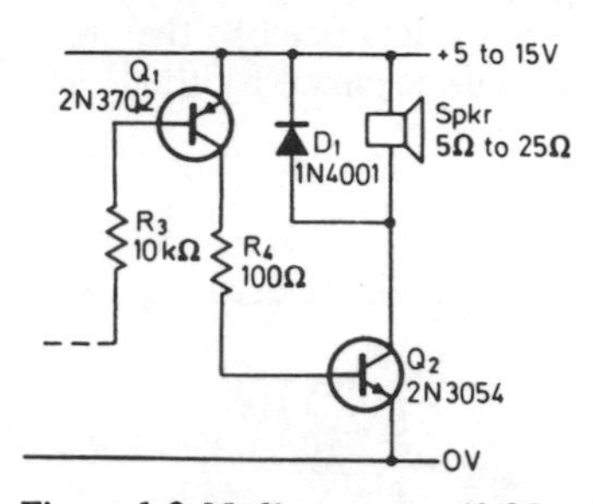

Figure 1.9 Medium-power (0.25–11.25 W) output stage

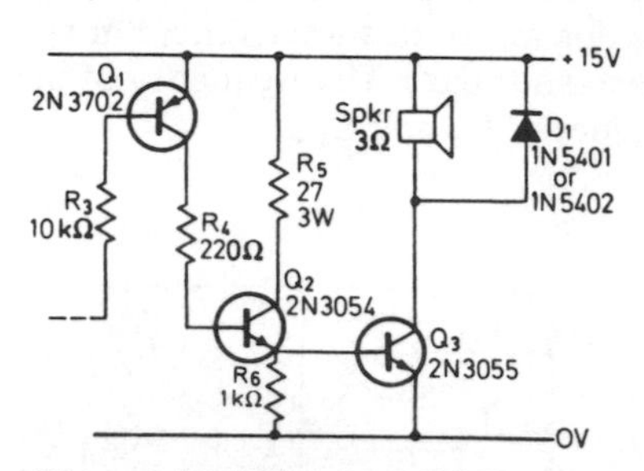

Figure 1.10 High-power (18 W) output stage

Figure 1.11 shows how a CD4001 IC can be used to make an alarm that generates a pulsed 800 Hz tone when its contacts are operated. Circuit operation is quite simple. The two left-hand gates of the IC are wired as a low-frequency (roughly 6 Hz) gated astable multivibrator that is activated via the contact switches, and the two right-hand gates

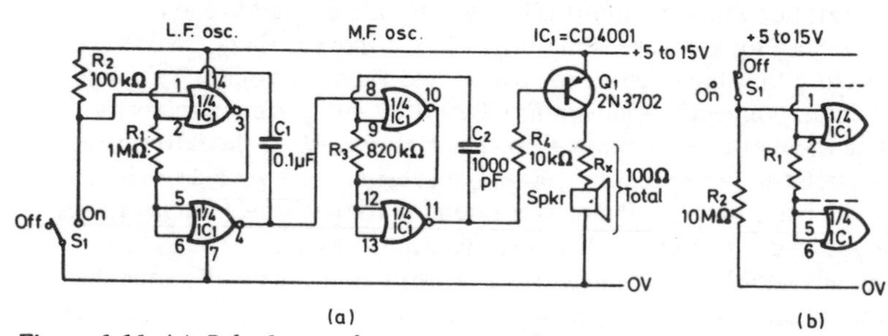

Figure 1.11 (a) Pulsed-tone close-to-operate alarm generator. (b) Modification for break-to-operate version

are wired as an 800 Hz astable multivibrator that is activated via the 6 Hz astable.

Normally, when the pin 1 input terminal of the circuit is high, both astables are inoperative and the circuit consumes only a small leakage current. When the input terminal is low, on the other hand, both astables are activated and the low-frequency circuit pulses the 800 Hz astable on and off at a rate of about 6 Hz, so a pulsed 800 Hz tone is generated in the speaker.

The *Figure 1.11* circuit can be activated via n.o. contact switches by using the input connections shown in *Figure 1.11a,* or by n.c. switches by using the connections shown in *Figure 1.11b.* If desired, the circuit's normal output power of only a few milliwatts can be boosted to as high as 18 W by replacing its Q_1 output stage by one or other of the *Figure 1.9* or *Figure 1.10* power-booster circuits.

Figure 1.12 shows how the *Figure 1.11* circuit can be modified so that it produces a warble-tone alarm signal. These two circuits are

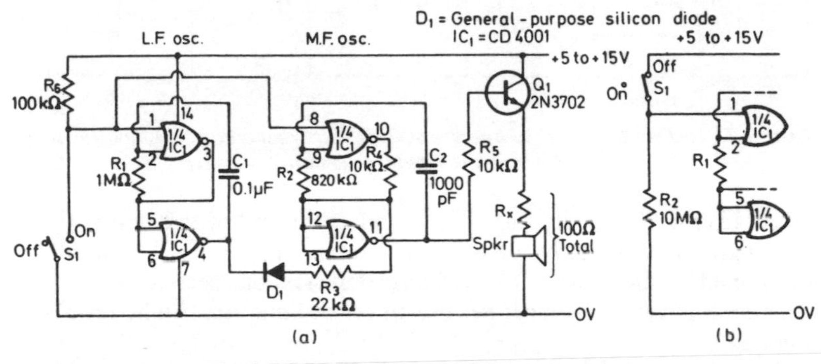

Figure 1.12 (a) Warble-tone close-to-operate alarm generator. (b) Modification for break-to-operate version

basically similar, but in the latter case the 6 Hz astable is used to modulate the frequency of the right-hand astable rather than to pulse it on and off. Note that the pin 1 and pin 8 gate terminals of both astables are tied together, and the astables are thus both activated directly by the contact switches. The circuit can be activated by n.o. switches by using the connections shown in *Figure 1.12a,* or by n.c. switches by using the connections shown in *Figure 1.12b.* The output power of the circuit can be boosted to a maximum of 18 W by using the power-booster stages of *Figure 1.9* or *1.10.*

The circuits of *Figures 1.8, 1.11* and *1.12* are all non-latching types, which produce an output only when they are activated by their contact switches. By contrast, *Figures 1.13* and *1.14* show two ways of using a CD4001 IC so that it gives some form of self-latch alarm-generating action.

The *Figure 1.13* circuit is that of a one-shot or auto-turn-off alarm generator. The action of this circuit is such that an 800 Hz monotone alarm signal is initiated as soon as contact switch S_1 is momentarily operated. This alarm signal then continues to be generated for a preset

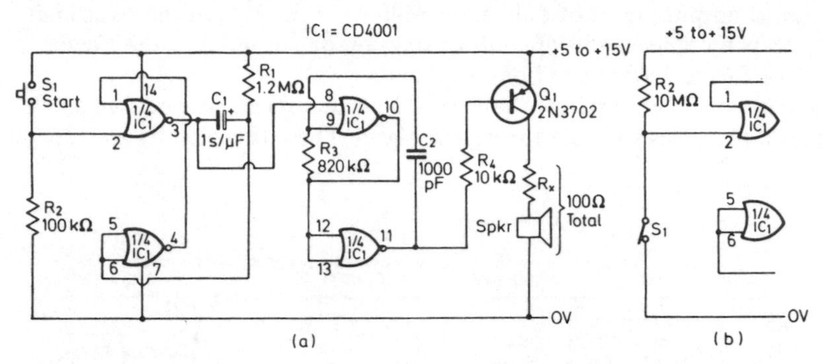

Figure 1.13 (a) One-shot close-to-operate 800 Hz alarm generator. (b) Modification for break-to-operate version

period, irrespective of the state of S_1, and at the end of this period the alarm signal automatically turns off. The duration of the alarm signal is determined by the value of C_1, and approximates one second per microfarad of value. Turn-off periods of several minutes can be readily obtained.

Here, the two left-hand gates of the IC are wired as a one-shot or monostable multivibrator, which can be triggered by a rising (positive-going) voltage on pin 2, and the two right-hand gates are wired as a gated 800 Hz astable multivibrator that is activated by the output of the monostable. Normally, both multivibrators are inoperative, and the circuit consumes only a small leakage current.

When S_1 is momentarily operated, a rising voltage is applied to pin 2 of the IC, and the monostable fires and gates the astable on, so an 800 Hz tone is generated in the speaker. At the end of the preset period the monostable automatically turns off again, so the tone ceases to be generated and the current consumption returns to leakage levels. The circuit can be reactivated again only by applying a rising voltage to pin 2 via S_1. The circuit can be activated by n.o. switches by using the connections shown in *Figure 1.13a,* or by n.c. switches by using the connections shown in *Figure 1.13b.*

Finally, *Figure 1.14* shows how the CD4001 can be used to make a true self-latching 800 Hz contact-operated alarm generator. Here, the two left-hand gates of the IC are wired as a manually-triggered bistable multivibrator, and the two right-hand gates are wired as a gated 800 Hz astable multivibrator that is activated via the bistable.

Circuit action is such that the output of the bistable is normally high, so the astable is disabled and the circuit consumes only a small leakage current. When S_1 is momentarily operated, a positive signal is applied to pin 2 of the IC, so the bistable changes state and its output locks into the low state and activates the astable multivibrator. An 800 Hz tone signal is generated in the speaker under this condition.

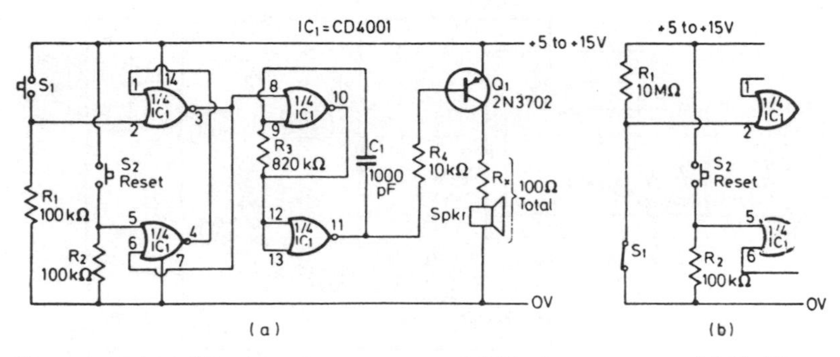

Figure 1.14 (a) Self-latching close-to-operate 800 Hz alarm generator. (b) Modification for break-to-operate version

Once it has been activated, the circuit can only be turned off again by removing the positive signal from pin 2 and briefly closing 'reset' switch S_2, at which point the circuit resets and its quiescent current returns to leakage levels.

The *Figure 1.14* circuit can be activated by n.o. switches by using the connections shown in *Figure 1.14a*, or by n.c. switches by using the connections shown in *Figure 1.14b*.

Note that the *Figure 1.13* and *1.14* circuits normally give output powers of only a few milliwatts, but that these levels can be boosted as high as 18 W by replacing their Q_1 output stages with the power-boosting circuits of *Figure 1.9* or *1.10*.

Multitone generator circuits

To conclude this chapter, *Figures 1.15* and *1.16* show the circuits of two push-button activated multitone alarm generators. These circuits have two or three sets of push-button operating switches, and the circuit action is such that each push switch causes the generation of its own distinctive tone. These circuits have uses in 'door announcing' applications, where, for example, a high tone may be generated by operating the front door switch, a low tone by operating the back door switch, and a medium tone by operating the side door switch.

The *Figure 1.15* circuit is that of a simple three-input monotone alarm generator system. Here, two of the gates of a CD4001 IC are wired as a modified astable multivibrator, and the action is such that

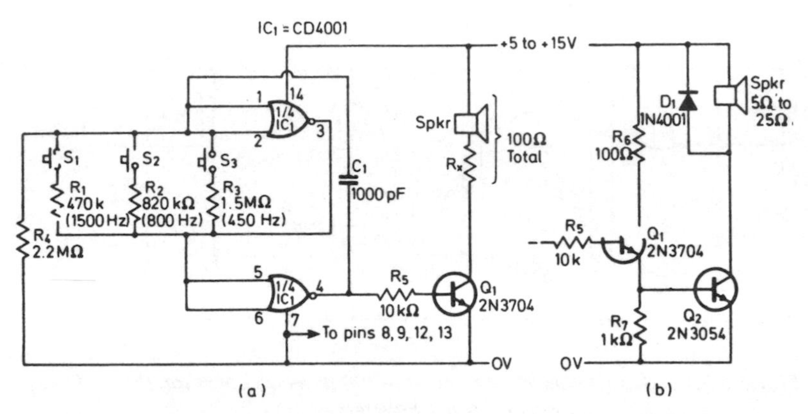

Figure 1.15 (a) Multitone contact alarm. (b) Modification for high-power output

the circuit is normally inoperative and drawing only a negligible leakage current, but becomes active and acts as a square-wave generator when a resistance is connected between pins 2 and 5 of the IC. This resistance must be substantially less than the 2.2 MΩ of R_4, and the frequency of the tone is inversely proportional to the resistance value that is used.

With the component values shown, the circuit generates a tone of roughly 1500 Hz when S_1 is operated, 800 Hz when S_2 is operated, and 450 Hz when S_3 is operated. Note that these tones are separated by roughly one octave each, so each push-button generates a very distinctive tone.

As in the case of the other circuits already described, the *Figure 1.15a* circuit generates an output power of only a few milliwatts. If required, this power can be boosted to as high as 11.25 W by using the power-booster stage shown in *Figure 1.15b.* As in the case of the *Figure 1.9* circuit, the final output power depends on the actual values of speaker impedance and supply rail voltage that are used.

Figure 1.16a shows the circuit of a two-input multitone circuit that generates a monotone signal when one push-button is operated, or a pulsed-tone signal when the other push-button is operated.

Here, the two left-hand gates of the IC are wired as a low-frequency (approximately 6 Hz) gated astable multivibrator, and the two right-hand gates are wired as a gated 800 Hz astable multivibrator. The two multivibrators are interconnected via silicon diode D_1, and the circuit action is such that the low-frequency astable oscillates and activates the 800 Hz astable when S_1 is operated, thus producing a pulsed tone

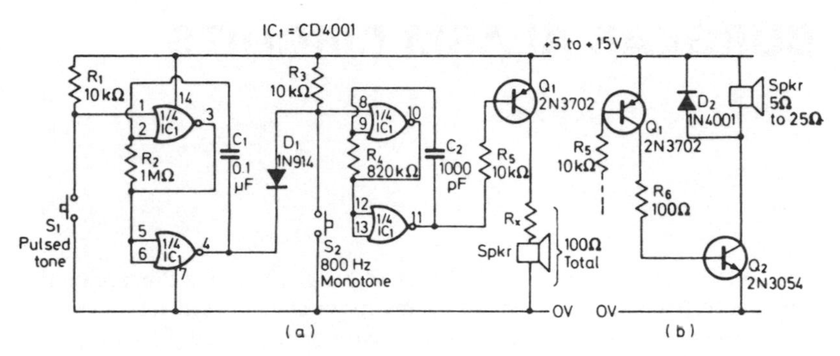

Figure 1.16 (a) Two-input multitone contact alarm. (b) Modification for high-power output

in the speaker, but only the right-hand 800 Hz astable operates when S_2 is closed, thus producing a monotone signal in the speaker.

The basic *Figure 1.16a* circuit generates an output power of only a few milliwatts. If required, this power can be boosted to as high as 11.25 W by using the power-booster stage shown in *Figure 1.16b*. The actual output power of this circuit again depends on the values of speaker impedance and supply-rail voltage that are used.

CHAPTER 2

BURGLAR ALARM CIRCUITS

The most reliable and widely-used type of burglar alarm is the 'contact' operated type, which is activated via microswitches or reed-relays wired into doors or windows, or via lengths of wire or foil wired into walls, floors or ceilings, etc. Ideally, such systems are battery powered, consume negligible standby current, provide a relay output for operating any type of alarm generator (bell, siren, etc.), and have optional provision for incorporating 'panic', 'fire', or similar alarm facilities.

In this chapter we show a variety of basic burglar-alarm and accessory circuits that meet the above ideals, and explain how the reader can combine different circuits to produce a 'tailor-made' alarm system that meets his own specific needs. These systems can be as simple or as complex as the individual reader cares to make them. We also show three 10 W alarm-call generator circuits that can be used in place of alarm bells or sirens, and give advice on how to install a complete alarm system in the home. Most of the alarm circuits described are designed around readily available COS/MOS digital integrated circuits.

Basic burglar-alarm circuits

Contact-operated burglar alarms can be designed as either self-latching circuits, which turn on as soon as they are activated and then remain on indefinitely (or until their supply batteries run down), or as auto-turn-off circuits, which turn on as soon as they are activated but then turn off again automatically after a preset period. Basic COS/MOS alarm systems of both these types are shown in *Figures 2.1* and *2.2.*

The self-latching operation of the *Figure 2.1* circuit is obtained by wiring two of the gates of CD4001 COS/MOS IC so that they act as a

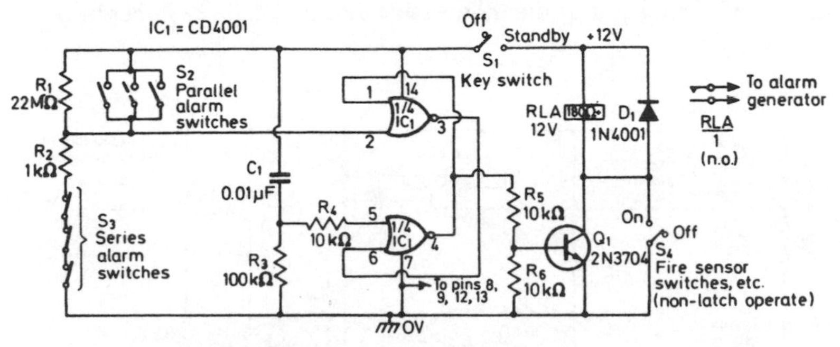

Figure 2.1 Simple self-latching burglar alarm

simple bistable multivibrator. The action of this bistable is such that its output (taken from pin 4) goes low and self-latches when a positive voltage or pulse is applied to pin 5, and its output goes high and self-latches when a positive voltage or pulse is applied to pin 2. Power is applied to the bistable and to the alarm sensor switches (S_2 and S_3) via key-operated switch S_1.

Assume, then, that S_2 is open and S_3 is closed. When key-switch S_1 is first set to the 'standby' position, pin 2 is held low by the potential divider action of R_1 and R_2, and a brief positive voltage pulse is fed to pin 5 from the supply line via C_1 and R_3–R_4. Consequently, the output of the bistable automatically goes low as soon as S_1 is closed. Under this condition zero base drive is applied to Q_1, so Q_1 and the relay and the alarm are all off. The circuit draws a typical current of about 1 μA in this standby mode: half of this current flows via R_1 and R_2, and the remainder as leakage via Q_1.

The alarm can be activated by opening any one of series-connected sensor switches S_3, or by closing any one of parallel-connected sensor switches S_2. Under this condition pin 2 of the bistable goes close to the positive supply-rail voltage, and the bistable changes mode and its output locks into the high state and switches the alarm generator on via Q_1 and the relay. The alarm then stays on indefinitely, and can be turned off only by opening S_1.

The auto-turn-off circuit of *Figure 2.2* is similar to that of *Figure 2.1,* except that the two gates of the IC are connected as a simple monostable multivibrator. The action of this monostable is such that its output goes to the low state when a positive voltage or pulse is fed to pin 5, but goes high for a preset period when a positive-going voltage transition is applied to pin 2. The value of this preset period is determined by the time constant of R_7 and C_2, and equals roughly four minutes (0.5 s/μF of C_2 value) with the C_2 value shown. At the end of this period the output of the monostable automatically switches back

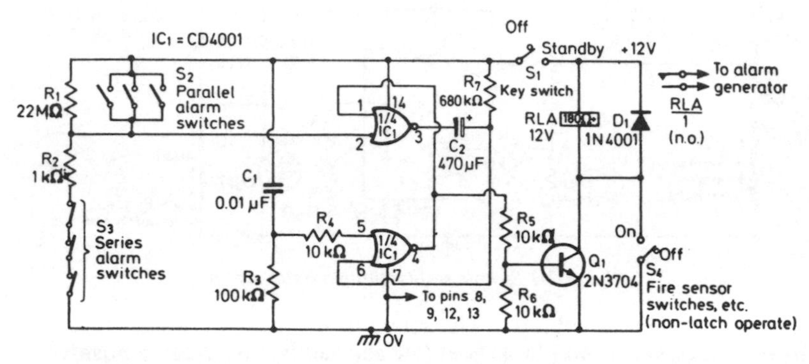

Figure 2.2 Simple auto-turn-off burglar alarm (turn-off delay ≈ 4 minutes)

to the low state. Note that the monostable can be triggered only by a positive-going transition of its pin 2 voltage, and its action is not influenced by 'standing' high or low voltages that may be applied to this pin.

Thus the output of the monostable automatically goes low as soon as key-switch S_1 is set to the 'standby' position. Under this condition the relay and the alarm are off, and the circuit consumes a typical standby current of 1 μA. The alarm can be activated by opening any one of series-connected sensor switches S_3, or by closing any one of parallel-connected sensor switches S_2. Under this condition a positive-going transition appears on pin 2 of the monostable, and its output switches into the high mode for a preset period and turns Q_1–RLA and the alarm on. At the end of this period the output of the monostable goes low again, irrespective of the states of S_2 and S_3, and Q_1–RLA and the alarm turn back off. The circuit can then be reset either by opening and then closing S_1, or by setting all S_2 and S_3 switches back to their original conditions.

Note in the *Figure 2.1* and *2.2* circuits that power is permanently applied to the Q_1–RLA sections of the designs, even when S_1 is in the 'off' position. This facility enables the alarm to be activated in the non-latching mode at all times via an n.o. temperature-sensing switch or thermostat, so that these circuits can also function as permanently-alert fire alarm systems. Any number of n.o. switches can be wired in parallel with S_4.

A weakness of the *Figure 2.1* and *2.2* circuits is that they give the owner no protection against intruders who may break into the house when the main alarm system is switched off. Protection against this type of intrusion can be obtained by scattering a number of series-connected n.c. 'panic' buttons around the house, so that a permanently alert self-latching alarm system can be activated manually at any time. This facility can readily be added to the *Figure 2.1* and *2.2* circuits, and *Figure 2.3* shows how it can be wired into the auto-turn-off system of *Figure 2.2*.

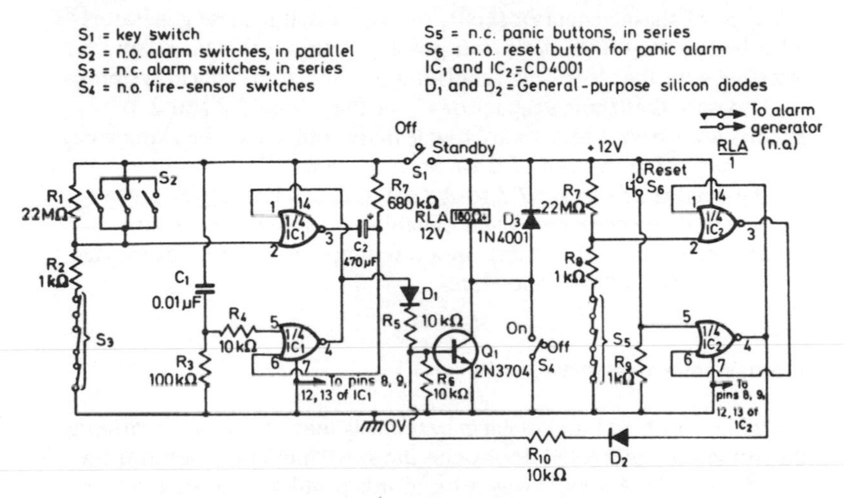

Figure 2.3 Auto-turn-off alarm with 'panic' facility

Here, part of IC_2 is wired as a simple bistable multivibrator that is permanently connected to the supply lines. The output of the bistable is taken to the base of Q_1 via D_2 and R_{10}, so that Q_1 and the relay and alarm can be turned on via the bistable. D_1 is wired in series with R_5 of the main alarm system so that the two sections of the circuit do not interact adversely. The output of the bistable is normally latched into the low state, so the relay and alarm are normally off. If any of series-connected 'panic' buttons S_5 are opened, the bistable immediately changes mode and its output locks into the high state and drives *RLA* and the alarm on. Once the alarm has been turned on, it can be reset to the 'off' state by briefly closing 'reset' switch S_6. This panic facility adds 0.5 μA to the total quiescent current consumption of the complete alarm system.

The panic facility can be added to the *Figure 2.1* circuit by simply wiring D_1 in series with R_5, and adding IC_2 and its associated circuitry to the basic design.

Note in the *Figure 2.3* circuit that two independent CD4001 ICs are used. This is because all four of the gates in each IC are connected to the same supply-line points (pins 7 and 14), and in *Figure 2.3* we need to be able to remove the supply from one pair of gates while keeping it on the other. All unused pins of the ICs are tied to ground, as indicated.

Note that the relay used in each of the *Figure 2.1* to *2.3* circuits can be any 12 V type with a coil resistance of 180 Ω or greater, and with one or more sets of n.o. contacts. The contacts can be used to activate any type of alarm generator (bells, sirens, etc.), but these generators must be operated from their own power supplies, otherwise they may interfere with the electronic functioning of the actual alarm systems.

Also note that timing capacitor C_2 of the *Figure 2.2* and *2.3* circuits must have a reasonably low leakage, otherwise the alarms may fail to turn off at the end of their preset periods.

The circuits of *Figures 2.1* to *2.3* act as excellent burglar-alarm systems in their own right. Their capabilities can be considerably expanded, however, by adding on a few simple electronic accessories, as shown in the following section.

Alarm-system accessory circuits

A problem with all burglar-alarm systems is that of leaving or entering the house via a protected door once the system has been set into the 'standby' mode. A simple way around this problem is to fit a key-operated by-pass switch to the outside of the door, so that the door's sensor switch can be temporarily disabled by the authorised key holder.

In this case the procedure for leaving the house is first to open the door and disable its sensor via the key switch, then re-enter the house and set the alarm to 'standby', and then leave the house again, close the door and re-enable its sensor via the key switch. The procedure for re-entering the house without sounding the alarm is simply to disable the door sensor via the key switch, then enter the house and turn the alarm system off.

Most of the tedium of this procedure can be eliminated by equipping the alarm system with an 'exit delay' facility, which automatically disables the door sensor for a preset period after the main alarm system is switched to 'standby'. This facility enables the owner simply to switch the alarm system to 'standby' and then leave the house without sounding the alarm, but it is still necessary for the owner to disable the door-sensor switch manually on re-entry if entry is to be made without sounding the alarm.

If required, even this re-entry procedure can be eliminated by equipping the alarm system with a combined 'exit and entry delay' facility. This ensures that the alarm will not sound until a preset time after the door sensor is initially activated by the entry action, thus giving the owner time to enter the house and turn off or reset the alarm system before the alarm actually sounds.

Practical 'exit delay' and 'exit and entry delay' circuits are shown in *Figures 2.4* and *2.5.* These facilities can readily be added to any of the main alarm-system circuits shown in *Figures 2.1* to *2.3.*

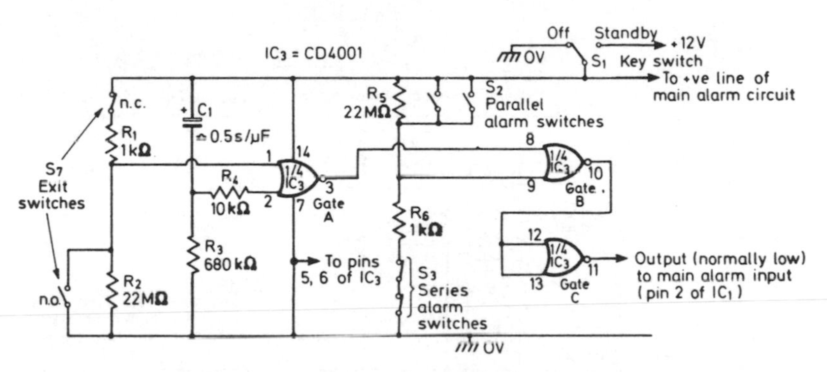

Figure 2.4 Alarm system 'exit delay' facility

The 'exit delay' facility of *Figure 2.4* uses three gates of a CD4001 IC. Door-sensor switch S_7 can be of either the n.o. or n.c. type, and is connected in such a way that the input to pin 1 of gate A is at positive rail voltage when the door is closed, and at ground volts when the door is open. Gate A is wired as a simple NOR gate, which gives a low output when either input is high, and time-delay network C_1–R_3 is connected to the pin 2 input of the gate via R_4. When power is first applied to the circuit C_1 is fully discharged, so pin 2 is effectively shorted to the positive supply line via R_4, and the output of the gate is at ground volts, irrespective of the state of the door sensor switch. After a delay determined by C_1 and R_3 (roughly 0.5 s/μF of C_1 value) the pin 2 voltage decays to such a value that the gate is influenced by the state of the door sensor switch. If the door is closed at this point the gate output remains low, but if the door is open the output goes high.

The output of gate A is taken directly to pin 8 of gate B, which is also connected as a NOR gate, and the main section of the alarm system's sensor circuitry is taken to pin 9 of this gate in such a way that this pin is effectively grounded under normal conditions. The output of gate B is inverted by gate C, which thus gives an output that is normally low, and this output is passed on directly to pin 2 of IC_1 in the main alarm circuit.

Thus, the action of the *Figure 2.4* circuit is such that all sensor switches except S_7 are enabled as soon as S_1 is set to the 'standby' position, and S_7 is disabled for a preset period. At the end of this period S_7 is enabled, and the alarm is able to respond to the actions of S_7.

The combined 'exit and entry delay' facility circuit of *Figure 2.5* is similar to that of *Figure 2.4*, except that R_1 is increased to 10 kΩ,

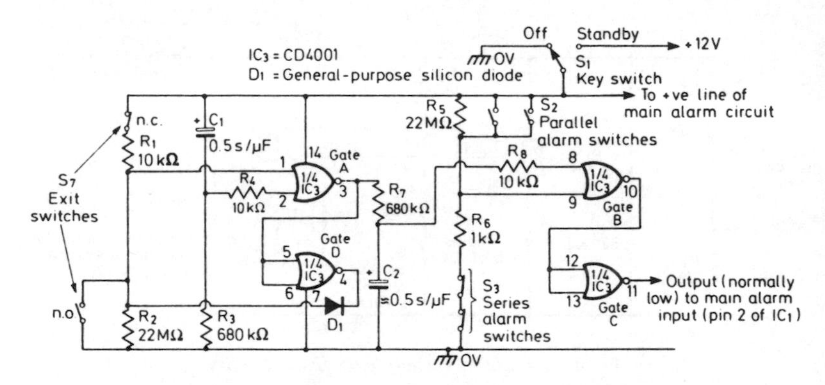

Figure 2.5 Alarm system 'exit and entry delay' facility

gate A is converted into a self-latching switch with the aid of D_1 and gate D, and the output of gate A is fed to the input of gate B via time-delay network C_2–R_7 and R_8. Circuit action is as follows.

When power is first applied to the circuit all sensor switches are enabled except S_7, which is disabled for a preset period via time-delay network C_1–R_3. The output of gate A is held in the low state under this condition. At the end of this preset period S_7 is enabled. If S_7 is activated after the end of this preset period, the output of gate A immediately goes high, and is locked in this state by the actions of D_1 and gate D. This high output voltage is applied to the input of gate B via time-delay network C_2–R_7, and after a preset delay (roughly equal to 0.5 s/μF of C_2 value) the voltage reaching gate B rises to such a value that the alarm is activated.

The circuit of *Figure 2.4* or *2.5* can be added to any of the main alarm circuits of *Figures 2.1* to *2.3* simply by removing the existing connections to pin 2 of IC_1, rewiring the existing alarm sensors into the *Figure 2.4* or *2.5* circuit, and connecting the outputs of this circuit to pin 2 of IC_1. Note that it is also necessary to wire the 'off' pin of key-switch S_1 to ground if a delay circuit is used, so as to provide a discharge path for its timing capacitors.

All the burglar alarm circuits shown in this chapter give reliable performances, and are not prone to giving false alarms under normal circumstances. One 'exceptional' circumstance which may initiate false alarms in any type of alarm system is a thunderstorm, where heavy electrical discharges may induce such large energy pulses into the alarm sensor wiring that the alarm is made to trigger falsely. In COS/MOS alarm systems this possibility can be eliminated by interposing 'sensor

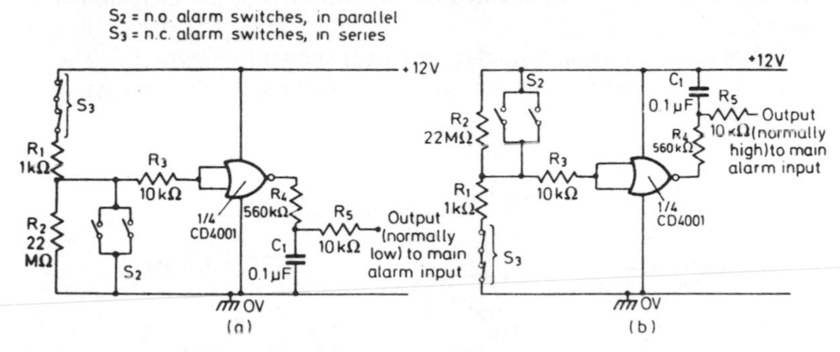

Figure 2.6 Sensor transient suppressors: (a) normally-low output; (b) normally-high output

transient suppressors' between the outputs of the main sensor networks and the inputs of the main alarm systems. *Figure 2.6* shows practical circuits of this type.

Here, a spare gate of a CD4001 IC is wired as a simple inverter, and the input of this gate is connected to the output of the main sensor network via limiting resistor R_3. The output of the gate is taken to the input of the main alarm via R_5 and time-constant network C_1-R_4. This network only passes signals that are applied to the gate input for periods greater than 50 ms. Consequently, the circuit rejects short-duration spurious pulses that are induced into the sensor wiring, but passes longer-duration signals that are generated by the activation of the sensor switches.

The *Figure 2.6a* circuit is intended for applications where the sensor input to the main alarm system is required to be normally low, and the *Figure 2.6b* circuit is for use where the sensor input needs to be normally high. In practice, these 'transient suppressor' circuits are only likely to be needed in cases where the lengths of alarm-sensor wiring exceed fifty metres or so, since all the alarm circuits shown in this chapter have fairly low input impedances (1 kΩ or 10 kΩ) when the sensor switches are in their normal states, and are thus not unduly sensitive to induced signals.

One final accessory that can be added to a burglar alarm system is an 'intrusion recorder'. This gadget is intended for use in auto-turn-off alarm systems only, and consists of a low-power sound generator that turns on and self-latches if an intrusion occurs, thus giving a continuous indication of the intrusion. The device can tell the owner that an intrusion has occurred during his absence from the house, even though the main alarm system has turned off and no signs of the intrusion are visible.

A practical 'intrusion recorder' circuit is shown in *Figure 2.7*. The circuit is permanently wired across the supply lines, and its operation

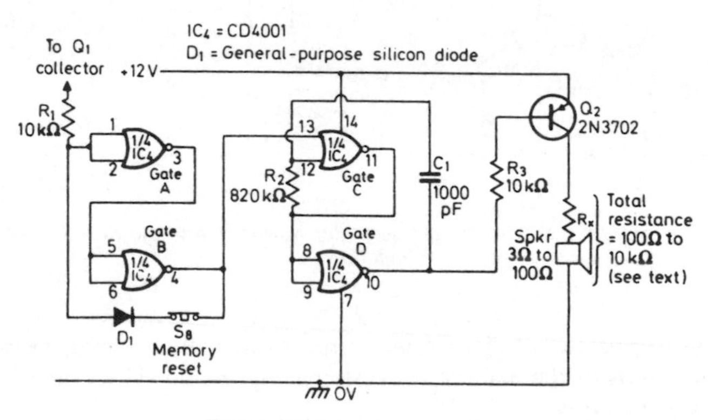

Figure 2.7 Intrusion recorder

is quite simple. Gates C and D are wired as a gated 800 Hz oscillator, which drives a speaker via Q_2 and R_X, and is activated from the collector of Q_1 of the main alarm system via the self-latch switch formed by gates A and B. Normally, the collector voltage of Q_1 is high and the alarm relay is off, and under this condition the 800 Hz oscillator is inoperative and the recorder circuit consumes a quiescent current of about 1 μA. If the main alarm system is activated the relay turns on for a preset period and the collector of Q_1 goes high. Under this condition, gates A and B of the recorder turn on and self-latch, and activate the 800 Hz oscillator, thus causing an audible signal to be generated in the speaker. Once this signal has been initiated, it can only be stopped again by operating 'reset' switch S_8.

The *Figure 2.7* circuit can be added to the auto-turn-off circuit of *Figure 2.2* or *2.3* simply by wiring it across the supply lines and connecting R_1 to the collector of Q_1. The speaker used in the circuit can have any impedance in the range 3 Ω to 100 Ω. The combined series value of R_X and the speaker impedance can be varied from a minimum value of 100 Ω up to 10 kΩ, depending on the sound level that is wanted from the speaker. The maximum power output of the circuit is about 250 mW when R_X has a value of zero and a 100 Ω speaker is used, and in this case the circuit consumes roughly 50 mA of current. Proportionately lower currents are consumed at lower power levels.

A comprehensive alarm system

The alarm system accessory circuits of *Figures 2.4* to *2.7* can be added to the basic alarm circuits of *Figures 2.1* to *2.3* in any combination, depending on the requirements of the individual reader. The final alarm system can be as simple or as complex as the reader desires.

The comprehensive alarm system of *Figure 2.8* is shown as an example of how a number of different circuits can be wired together to meet a specific alarm system requirement. In this case the alarm is of the auto-turn-off type, has a 'panic' facility and an intrusion recorder, and is intended for use with an n.o. exit/entry switch. The system incorporates an 'exit and entry delay' facility, giving delays of about 25 seconds in each mode, and has transient suppression applied to the main sensor network. The system has provision for non-latch activation via n.o. heat-sensing switches, and thus also functions as an automatic fire alarm.

The 'panic' facility is designed around IC_2, and the intrusion recorder is designed around IC_4. Both of these sections of the circuit are permanently wired across the supply lines. The auto-turn-off operation is obtained via IC_1, and one of the spare gates of this IC is used to provide transient suppression for the main sensor network. Finally, IC_3 provides the 'exit and entry delay' facility. All four of the ICs used in the system are CD4001 types. Note that the 'off' terminal of key-switch S_1 is taken directly to ground, to provide a discharge path for the system's timing capacitors.

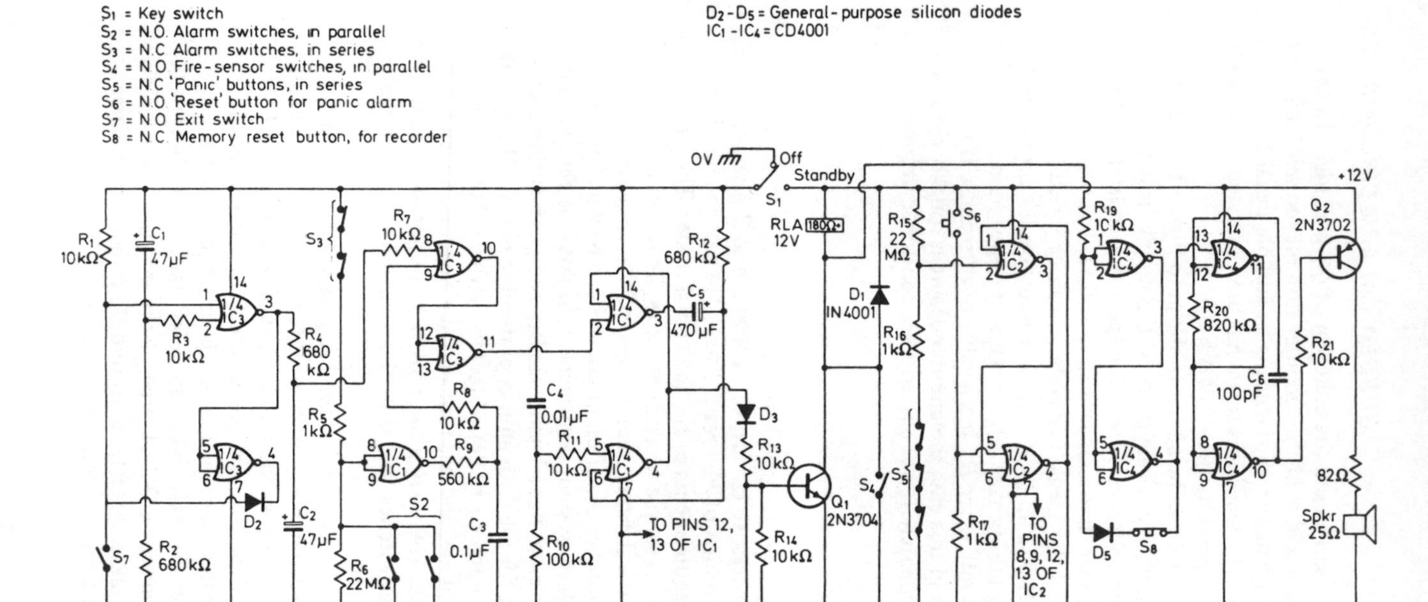

Figure 2.8 Comprehensive alarm system

A utility burglar alarm system

The burglar alarm circuits shown so far in this chapter are presented with the aim of enabling the reader to build an alarm system that meets his own specific requirements, which may be very simple or quite complex. By contrast, *Figure 2.9* shows the circuit of a utility burglar alarm that gives a very useful but restricted performance in home applications. Briefly, this circuit is designed for use with series-connected n.c. switches, and the alarm gives self-latching operation, so that once it is activated it continues to sound until it is turned off via a key-switch or its supply batteries run flat. The circuit action is such that an LED (light-emitting diode) illuminates if any of the n.c. sensor switches is open, but the actual alarm generator is automatically disabled for about 50 seconds when the circuit is first set to 'standby' via its key-switch. The actual alarm generator can use the same power supply as the *Figure 2.9* circuit, and the complete system can readily be modified to give 'fire' and 'panic' facilities.

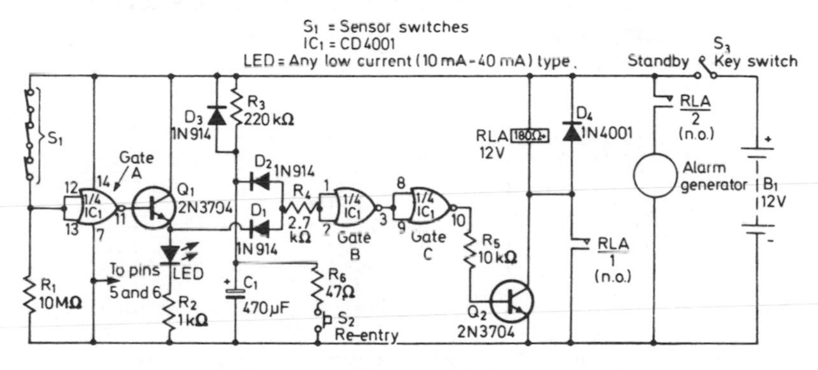

Figure 2.9 Utility burglar alarm system

In use, the alarm is first set to 'standby' via the key-switch, and the owner can then check by the LED that all sensor switches are correctly closed. The owner then has about 50 seconds in which he can leave the house without sounding the alarm. At the end of this period the whole alarm circuit becomes enabled, and if any of the sensor switches is subsequently opened the alarm generator activates and self-latches until the system is turned off manually or the supply batteries run flat. The owner can enter the house without activating the alarm by operating a simple push-button disabling switch fitted to the outside of the entry door.

The operation of the *Figure 2.9* circuit is quite simple. The series-connected n.c. sensor switches are taken to the input of gate A, which is used as a simple inverter, and the output of the inverter is taken to the LED via emitter-follower Q_1. The output of Q_1 is fed to relay-driving transistor Q_2 via a two-input diode AND gate formed by D_1 and D_2, and via gates B and C, which are connected as a non-inverting buffer stage. The other input of the AND gate is taken from the junction of the R_3-C_1 time-delay network; the action here is such that C_1 is fully discharged when power is first applied to the circuit, so gates B and C and transistor Q_2 are disabled, but after a delay of about 50 seconds C_1 charges to such a level that these components become enabled, and the relay operates if any of the input switches open and the LED goes on. The relay is made self-latching via the n.o. *RLA*/1 contacts, and the n.o. *RLA*/2 contacts are used to operate the alarm generator.

Note that the circuit can at any time be disabled for a period of about 50 seconds by briefly operating push-button switch S_2, so that it discharges C_1. In practice, S_2 can be concealed outside the house near the main entry point, so that the owner can enter the house without sounding the alarm. D_3 automatically discharges C_1 when power is removed from the circuit.

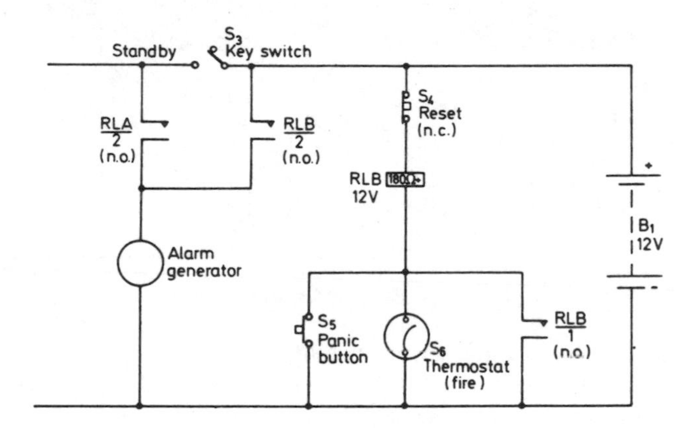

Figure 2.10 Modification to Figure 2.9 circuit: 'fire' and 'panic' protection fitted

Finally, *Figure 2.10* shows how an additional relay can be wired into the above circuit so that it also gives 'fire' and 'panic' protection. Here, n.c. push-button switch S_4 and the parallel-connected combination $S_5-S_6-RLB/1$ are all wired in series with relay B, and the combination is permanently wired across the supply lines. The n.o. *RLB*/2 contacts are used to connect the alarm generator to the supply lines.

The circuit action is such that *RLB* is normally off, but the relay turns on and self-latches and activates the alarm generator if any of the S_5 or S_6 switches briefly close. Any number of n.o. 'panic' buttons can be wired in parallel with S_5, and any number of n.o. thermostats can be wired in parallel with S_6. Once *RLB* has turned on, it can be turned off again by briefly opening S_4, which may be located in a concealed position.

10 W alarm-call generator circuits

All the burglar alarm circuits shown in this chapter give relay outputs, which can be used to activate any type of alarm-call generator. They can be used to activate bells, sirens, or electronic generators that produce an output in a loudspeaker. Three suitable 10 W electronic generators are described in this section.

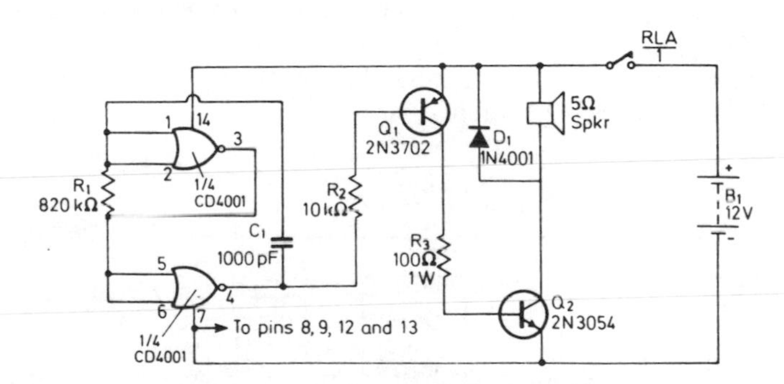

Figure 2.11 Monotone 10 W alarm-call generator

Figure 2.11 shows the circuit of a simple 10 W monotone alarm-call generator. Here, two gates of a CD4001 IC are inter-connected as an 800 Hz square-wave generator. The output of the generator is fed to a 5 Ω speaker via a direct-coupled power amplifier stage formed by Q_1 and Q_2. The action of the circuit is such that the transistors are alternately switched from the fully off to the saturated state at a rate of 800 Hz, so the power losses of the circuit are low. More than 10 W of power are fed to the speaker from the 12 V supply. Note that the two unused gates of the IC are disabled by wiring their input pins (pins 8, 9, 12 and 13) to pin 7.

Figure 2.12 shows the circuit of a pulsed-tone alarm-call generator, which produces an 800 Hz tone that is pulsed on and off at a rate of

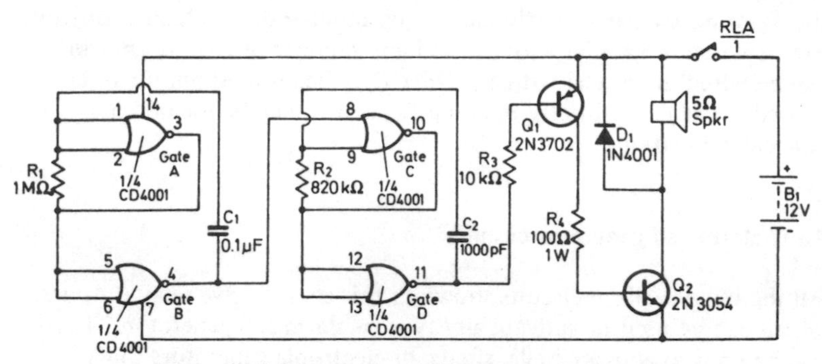

Figure 2.12 Pulsed-output 10 W alarm-call generator

6 Hz. Here, gates A and B are wired as a 6 Hz square-wave generator, which is used to alternately enable and disable the 800 Hz oscillator formed by gates C and D. The output of the 800 Hz oscillator is fed to the speaker via Q_1 and Q_2, and more than 10 W of power are fed to the speaker from the 12 V supply.

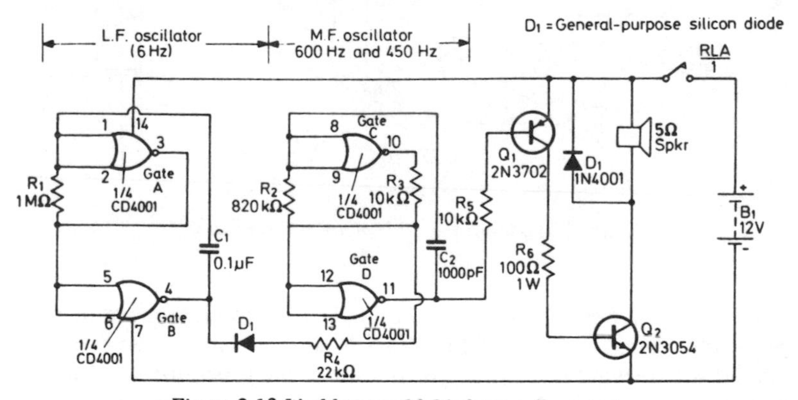

Figure 2.13 Warble-tone 10 W alarm-call generator

Finally, *Figure 2.13* shows the connections for making a warble-tone generator, in which the output switches alternately between 600 Hz and 450 Hz at a rate of 6 Hz. Here, the 6 Hz oscillator formed by gates

A and B is used to vary the periods and thus the frequency of the oscillator formed by gates C and D. The output of the IC is fed to the speaker via Q_1 and Q_2, and the output power of the circuit is greater than 10 W.

Note that the alarm-call generator circuits of *Figures 2.11* to *2.13* each use a 12 V battery supply. Also note that each circuit uses a 5 Ω speaker, and that a damping diode is wired across this speaker to suppress unwanted back e.m.f.s.

Each alarm-call generator circuit can be activated from the main alarm system by wiring the alarm's n.o. *RLA* contacts in series with the generator's positive supply lines, as shown in the diagrams. Note that, except in the case of *Figure 2.9,* the generator must use supplies that are independent of those of the main alarm system.

Alarm sensor systems

All the alarm circuits described in this chapter are 'contact operated' types. They are activated by the making or breaking of electrical contacts that are built into simple 'sensor' devices. These sensors can take the form of microswitches or reed-relays that are activated by the opening of a door or window, or of pressure pads that close when a person treads on a rug or carpet, or of lengths of wire or foil that break when a person forces an entry through a window, wall, floor or ceiling.

The selection of a complete alarm sensor installation depends on a number of factors. Amongst these are the physical details of the building that is to be protected, the value of the goods that are to be protected, and the ideas on crime prevention of the individual property owner. The choice of an installation is a very personal matter; the following notes are given to help the reader make that choice.

Any building can, for crime prevention purposes, be regarded as a box that forms an enclosing perimeter around a number of interconnected compartments. This perimeter 'box' is the shell of the building, and contains walls, floors, ceilings, doors and windows. To commit any crime within the building, the intruder must break through this perimeter, which thus forms the owner's first line of defence.

Once an intruder has entered the building, he can move from one room or 'compartment' to the next only along paths that are predetermined by the layout of internal doors and passages. In moving from one compartment to the next he must inevitably pass over certain 'spots' in the building, as is made clear in *Figure 2.14,* which shows the ground-floor plan of a small house. Thus to move between the lounge and the hall he must pass over spot X_1, to move between the kitchen and the hall he would tend to pass over spot X_2, and to move from the ground floor to the upper floor he must pass over spot X_3. These 'spot' points form the owner's second line of defence.

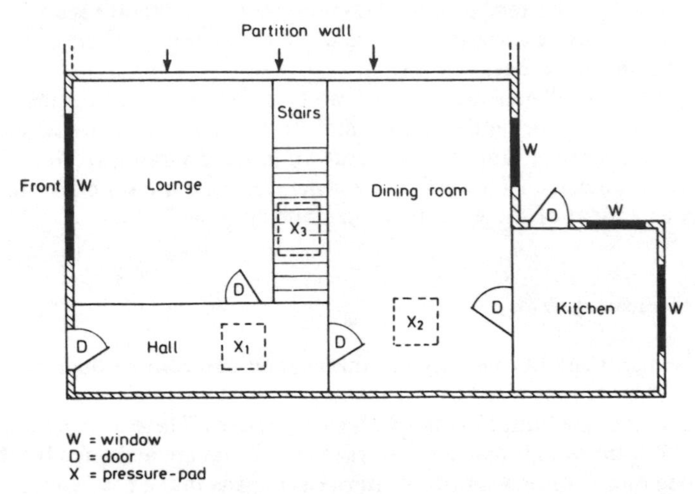

Figure 2.14 Ground-floor plan of small house, showing suitable positions for pressure-pad 'spot' defences

Thus the owner can obtain protection by using full or partial 'perimeter' defence, or by using 'spot' defence, or by using a combination of the two methods.

The most expensive type of alarm-sensor installation that can be fitted is the full perimeter defence system, which includes series-connected sensor wires built into all walls, floors and ceilings, as well as microswitches or reed-relays on all doors and windows. This type of installation is normally fitted only to commercial buildings such as jewelry stores or lock-up shops, etc., where the risk of burglary by skilful intruders is very high.

The least expensive type of alarm-sensor installation is the spot defence system, which can consist of just two or three pressure pads wired in parallel and hidden under rugs or carpets. This type of installation is adequate where the risk of burglary is small and the value of the protected goods is fairly low.

Intermediately priced partial perimeter defence installations can range from something as simple as a microswitch on a single side or rear door, to something that includes microswitches or reed-relays on all doors and window frames, plus protective foil on all windows and skylights. These systems can give adequate protection against most amateur and professional burglars, particularly when the installation is coupled to a spot defence system.

Burglars can, in general terms, be described as being of three basic types. The most common is the novice or amateur burglar, who enters a house at random in the hope of finding items worth stealing. This type of intruder usually has insufficient skill or motivation to beat even the simplest detector devices, and will flee at the first sound of an alarm bell.

The second type of intruder is the small-time professional. This type of burglar breaks into a house only if he is sure that it contains valuable items. Before attempting to enter a house he makes a thorough reconnaissance of its defence systems, and commits the actual burglary only if he thinks he has found an unprotected entry point, such as a skylight or an accessible ceiling or floor. He may be so 'cool' that he will ignore an alarm bell for several minutes before fleeing. The best defence against this type of intruder is a carefully thought out partial perimeter system, combined with a few 'spot' defence points.

Finally, the most difficult burglar to beat is the organised or gang professional, who plays for high stakes and will go to great lengths to win. He may be willing to simply crash his way through a defence wall, or hurt anyone that gets in his way. He may be undeterred by the sound of an alarm. The most effective defence against this type of criminal is a multiple perimeter system, in which the main building is surrounded by a partially protected outer perimeter, such as a wall, and all valuables are held within a fully protected inner perimeter, such as a strong-room.

Note that all alarm systems should, ideally, be fitted with a panic facility, to enable the owner to summon aid if an intrusion occurs while he is on the premises.

Different crime-prevention authorities have different ideas on the best way to protect a home against burglary. Some claim that every effort should be made to keep burglars out of the house at the outset, and that all possible points of entry should be protected. Others claim that a determined and skilful burglar can get past all but the most comprehensive of perimeter defence systems, so the most sensible approach is to have a very simple partial perimeter defence system combined with a good spot defence network, so that an intruder can enter the premises with relative ease but is scared off as soon as he gets inside.

Thus there are many points to consider when selecting a sensor system, and the reader must make up his own mind as to the best system to use in his particular case. Once the sensor system has been selected, the layout of the full alarm system installation must be considered. The following notes should be of value in this respect.

Alarm system installations

Figure 2.15 shows how a full alarm system installation can be broken down into three basic 'blocks', namely the sensor network, a control centre, and the alarm-call generator. The layout of the sensor network has already been discussed, and is a matter for individual decision.

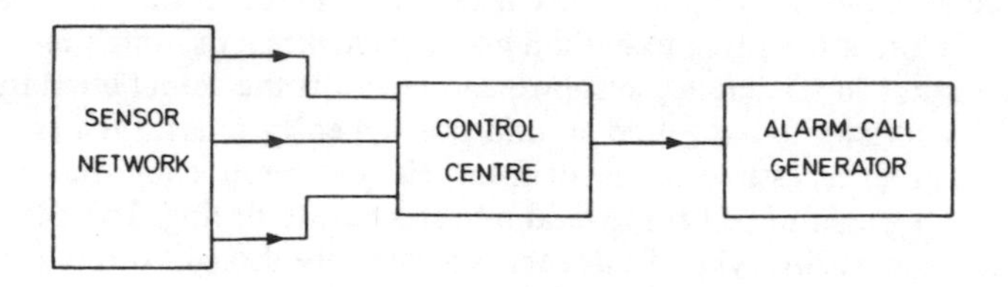

Figure 2.15 Block diagram of practical alarm-system installation

The alarm-call generator can be mounted in a prominent position on the front of the building, to act as a deterrent to would-be burglars, or it can be concealed in the eaves of the house in such a position that it can be heard equally well inside and outside the building. In either instance, the generator and its battery supply should be housed in a strong, burglar-proof box, and connected to the control centre either via an armoured cable or via cable that is concealed in the plasterwork, etc.

The control centre contains the electronics of the alarm system, together with the system's supply battery, plus a number of switches that enable different parts of the system to be turned on or off or to be tested. The centre should ideally be housed in a burglar-proof box, and the connections to the sensors should be made via armoured cable or concealed wiring.

Figure 2.16 shows a typical control-centre instrument panel, with five control switches. Switch S_1 is the main alarm system's 'on/off' control. As mentioned earlier, certain sections of the alarm system (such as fire sensors and panic facilities) must be permanently enabled, so S_1 controls the burglar alarm section of the circuit only. S_2 enables any auxiliary sensor devices, such as flood, overheat or power-failure detectors, to be switched in or out of the alarm system. Switches S_3 to

S_5 enable individual sections of the burglar-alarm sensor system, such as front door, stair or garage defences, to be connected or disconnected from the circuit.

Front door defences	Stair defences	Garage defences	Main alarm system	Auxiliary inputs
ON	ON	ON	ON	ON
OFF	OFF	OFF	OFF	OFF
S_3	S_4	S_5	S_1	S_2

Figure 2.16 Typical control-centre instrument panel

Finally, *Figure 2.17* shows the connections for turning individual sections of the alarm sensor network on and off. Series-connected n.c. sensor networks can be enabled or disabled by wiring them in parallel with S_1, as shown in *Figure 2.17a*. The sensors are enabled when S_1 is

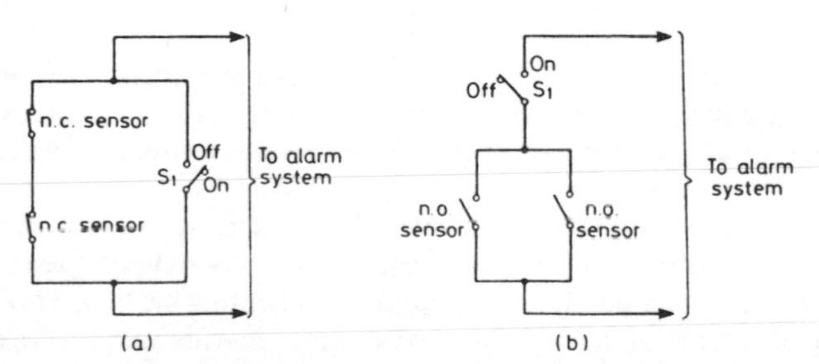

Figure 2.17 Method of enabling and disabling sensors via S_1 *: (a) series-connected, n.c., (b) parallel-connected, n.o.*

open, and are disabled when S_1 is closed. Parallel-connected n.o. sensor networks can be enabled and disabled by wiring them in series with S_1, as shown in *Figure 2.17b*. The sensors are enabled when S_1 is closed, and are disabled when S_1 is open.

CHAPTER 3

TEMPERATURE-OPERATED ALARM CIRCUITS

Temperature-operated alarms can be used as automatic fire or overheat alarms, as frost or underheat alarms, or as differential temperature alarms that operate when two temperatures differ by more than a preset amount. The alarms may be designed to give an audible loudspeaker output, an alarm-bell output, or a relay output that can be used to operate any kind of audible or visual warning device. The alarms may use thermostats, thermistors or solid-state devices as their temperature-sensing elements.

Temperature-operated alarms have many practical applications in the home and in industry. They can be used to give warning of fire, frost, excessive boiler temperature, the failure of a heating system, or overheating of a piece of machinery or of a liquid. A wide range of useful alarm types are described in this chapter.

Thermostat fire-alarm circuits

One of the simplest types of temperature-operated alarm is the thermostat-activated fire alarm. *Figure 3.1* shows the practical circuit of a relay-aided non-latching alarm of this type. Here, a number of n.o. thermostats are wired in parallel and then connected in series with the coil of a relay, and one set of the relay's n.o. contacts are wired in series with the alarm bell so that the bell operates when the relay turns on.

Normally, the thermostats are all open, so the relay and the alarm bell are off. Under this condition the circuit consumes zero standby current. At 'overheat' temperatures, on the other hand, one or more of the thermostats closes, and thus turns on the relay and thence the alarm bell. Note that push-button switch S_1 is wired in parallel with the thermostats, so that the circuit can be functionally tested by operating the push-button.

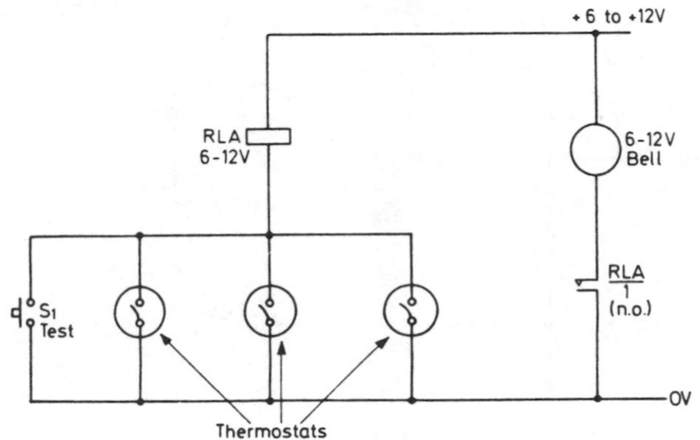

Figure 3.1 Simple relay-aided non-latching fire alarm

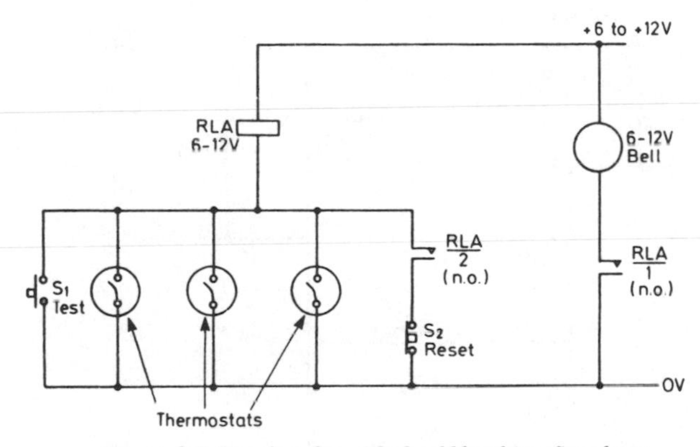

Figure 3.2 Simple relay-aided self-latching fire alarm

The thermostats used in this and all other circuits described here must be n.o. types that close when the temperature exceeds a preset limit. When the thermostats are located in normal living areas they should be set to close at a temperature of roughly 60°C (140°F), but when they are located in unusually warm places, such as furnace rooms or attics, they should be set to close at about 90°C (194°F).

The basic *Figure 3.1* circuit gives a non-latching form of operation. If required, the circuit can be made self-latching by wiring a spare set of n.o. relay contacts in parallel with the thermostats, as shown in *Figure 3.2*. Note that n.c. push-button switch S_2 is wired in series with the relay contacts, so that the circuit can be reset or unlatched by momentarily operating S_2.

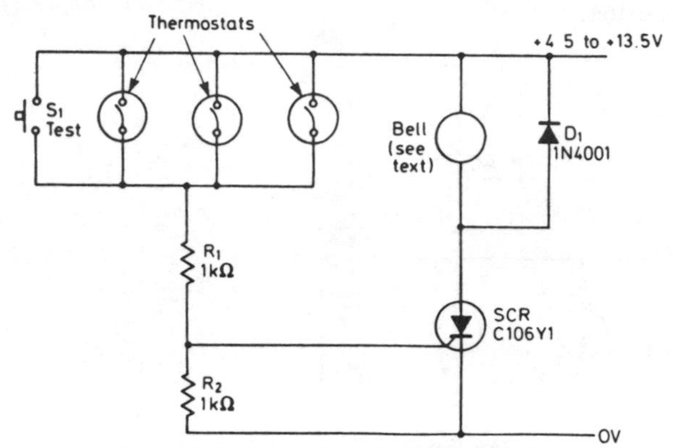

Figure 3.3 SCR-aided non-latching fire alarm

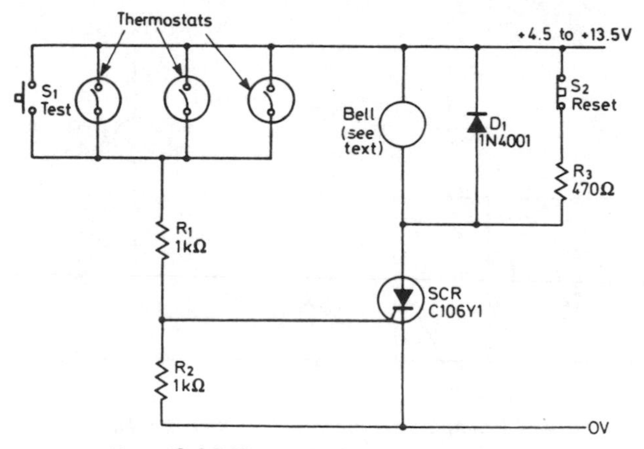

Figure 3.4 SCR-aided self-latching fire alarm

Bell-output fire-alarms can sometimes be activated via an SCR rather than a relay; *Figure 3.3* shows a typical circuit of this type. Here, a self-interrupting type of alarm bell is wired in series with the SCR anode, and gate current is provided from the positive supply line via the thermostats and via current-limiting resistor R_1. Normally the thermostats are open, so the SCR and the bell are off, and the circuit passes only a small leakage current. At high temperatures the thermostats close, so gate current is applied via R_1, and the SCR and the alarm bell turn on.

The basic *Figure 3.3* circuit gives a non-latching form of operation. The circuit can be made self-latching by wiring shunt resistor R_3 across the bell as shown in *Figure 3.4,* so that the SCR current does not fall below its latching value when the bell goes into its self-interrupting mode. Note that push-button switch S_2 is wired in series with R_3 so that the circuit can be reset or unlatched.

It should be noted that the SCR used in the *Figure 3.3* and *3.4* circuits has a current rating of only 2 A, so the alarm bell should be selected with this point in mind. Alternatively, SCRs with higher current ratings can be used in place of the device shown, but this modification may also necessitate changes in the R_1 and R_3 values of the circuits.

Thermostat fire alarms can be made to generate an alarm signal directly into a loudspeaker by using the connections of *Figure 3.5* or *3.6.* The *Figure 3.5* circuit generates a pulsed-tone non-latching alarm signal, while the *Figure 3.6* circuit generates an 800 Hz (monotone) self-latching alarm signal. Both are designed around a CD4001 COS/MOS digital IC.

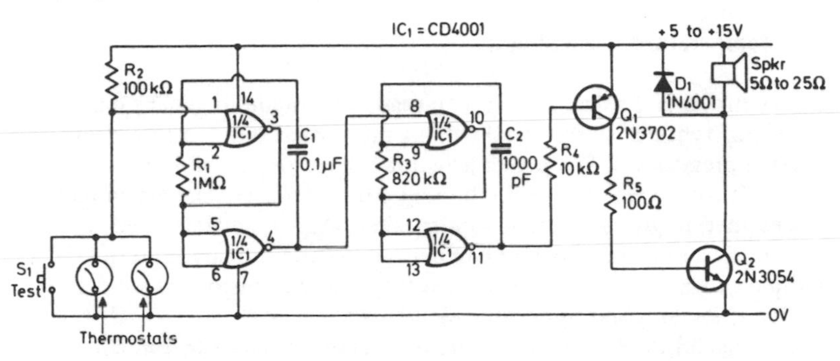

Figure 3.5 800 Hz pulsed-output non-latching fire alarm

The *Figure 3.5* circuit, which gives an output tone of 800 Hz pulsed on and off at a rate of 6 Hz, is based on the circuit of *Figure 1.11a* combined with a medium-power output stage. A full description of the circuit operation is given in Chapter 1.

The *Figure 3.6* self-latching circuit, which gives an 800 Hz monotone output, is based on the circuit of *Figure 1.14a* combined with a medium-power output stage. A full description of the operation of this circuit is also given in Chapter 1.

The circuits of *Figures 3.5* and *3.6* can be used with any supply voltages in the range 5–15 V, and with any speaker impedances in the range 5–25 Ω. The actual output power of each circuit depends on the

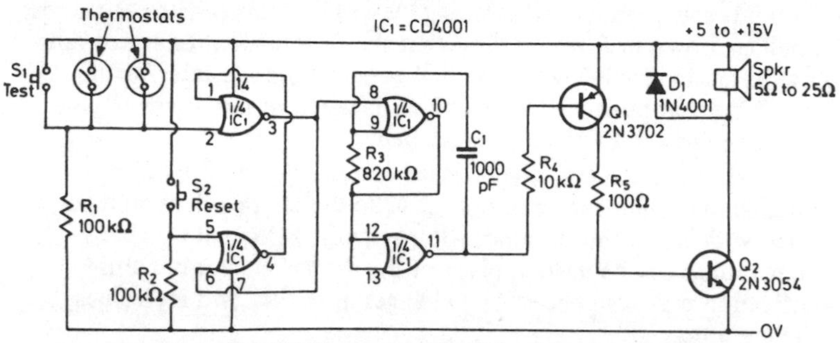

Figure 3.6 800 Hz monotone self-latching fire alarm

values of supply voltage and speaker impedance that are used, and varies from 0.25 W when a 25 Ω speaker is used with a 5 V supply, to 11.25 W when a 5 Ω speaker is used with a 15 V supply.

Over-temperature alarm circuits

Over-temperature alarm circuits can readily be made to generate a variety of types of alarm signal when a monitored temperature rises above a preset level. The preset level may range from well below Arctic temperatures to well above the boiling point of water. The alarm may be designed to give an audible loudspeaker output, an alarm-bell output, or a relay output that can be used to operate any kind of audible or visual warning device, and may use thermistors or solid-state diodes as its temperature-sensing elements.

Five useful over-temperature alarm circuits are described in this section. Most of them use inexpensive negative temperature coefficient (n.t.c.) thermistors as their temperature-sensing elements. These devices act as temperature-sensitive resistors that present a high resistance at low temperatures and a low resistance at high temperatures.

The thermistor circuits described in this and the following sections of this chapter have all been designed to work with thermistors that present a resistance of roughly 5 kΩ at the desired operating temperature. All these circuits are highly versatile, however, and will work well with any n.t.c. thermistors that present a resistance in the range 1 kΩ to 20 kΩ at the required temperature.

Figure 3.7 shows the practical circuit of a simple but highly efficient over-temperature alarm that gives a relay output. Here, the thermistor and R_1–R_2–R_3 are wired in the form of a simple bridge, in which R_1 is adjusted so that the bridge is almost balanced at the desired operating temperature, and a type 741 operational amplifier and transistor Q_1 are used as the bridge balance detector and relay driver.

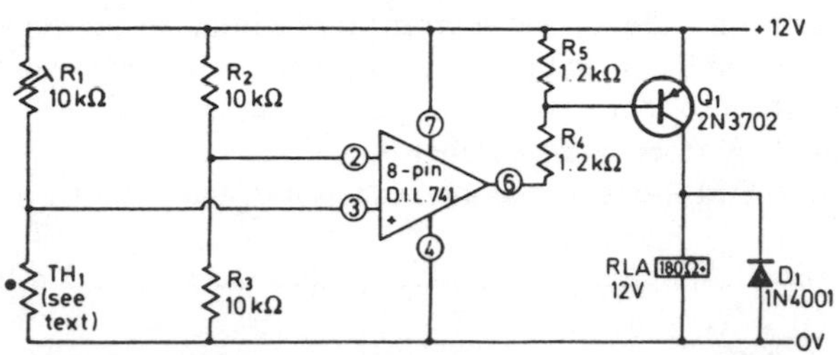

Figure 3.7 Relay-output precision over-temperature alarm

The 741 op-amp is used in the open-loop mode in the *Figure 3.7* circuit, and the device characteristics under this condition are such that its output (at pin 6) is driven to negative saturation (i.e. goes to almost zero volts) if its inverting (pin 2) input is more than a few millivolts positive to the non-inverting (pin 3) input, and is driven to positive saturation (i.e. goes to almost full positive-rail voltage) if its inverting input is more than a few millivolts negative to the non-inverting input. Thus when the bridge of the *Figure 3.7* circuit is close to balance, the op-amp can be driven from full positive saturation to full negative saturation, and vice versa, by pin 2 to pin 3 differential voltage changes of only a few millivolts.

Suppose, then, that the bridge is adjusted so that it is close to balance at the desired 'alarm' temperature. When the temperature falls below this value, the resistance of the thermistor increases, so the voltage on pin 3 of the op-amp rises above that of pin 2. Consequently, since pin 2 is negative to pin 3, the op-amp goes to positive saturation and applies zero base drive to Q_1, so Q_1 and the relay are off under this condition.

When, on the other hand, the temperature rises above the preset 'alarm' value, the resistance of the thermistor falls and the voltage on pin 3 of the op-amp falls. Consequently, since pin 2 is positive to pin 3, the op-amp goes to negative saturation and applies heavy base drive to Q_1, so Q_1 and the relay are driven on under this condition. Thus the relay goes on when the temperature rises above the preset level, and turns off when the temperature falls below the preset level.

Important points to note about the *Figure 3.7* circuit are that, because of the bridge configuration used, its accuracy is independent of variations in supply voltage, and that the alarm is capable of responding to resistance changes of less than 0.1 per cent in the thermistor, i.e. to temperature changes of a fraction of a degree.

Another point to note is that the circuit is quite versatile. It can, for example, be converted to a precision under-temperature alarm by simply transposing the R_1 and thermistor positions, or by transposing the pin 2 and pin 3 connections of the op-amp, or by redesigning the Q_1 output stage so that it uses an npn transistor in place of the pnp device. Similarly, there are a number of alternative ways of connecting the circuit so that it operates as a precision over-temperature alarm.

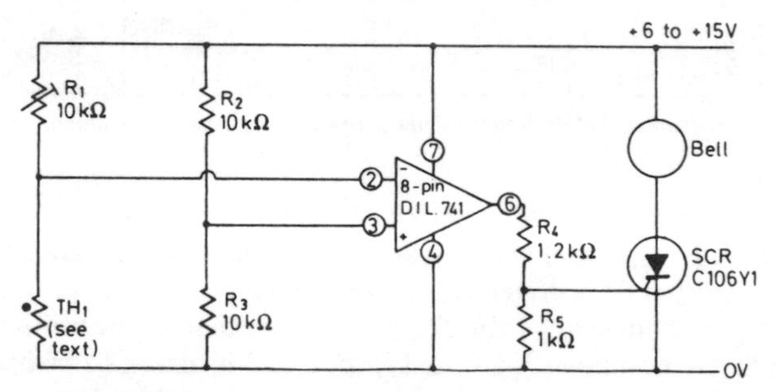

Figure 3.8 Direct-output precision over-temperature alarm

One such alternative, shown in *Figure 3.8,* provides a precision over-temperature alarm with an alarm-bell output. The circuit is similar to that of *Figure 3.7,* except that the pin 2 and pin 3 connections of the op-amp are transposed, and the output of the op-amp is used to drive the gate of an SCR rather than the base of a pnp transistor. The circuit action is such that the op-amp goes to negative saturation at 'low' temperatures, so zero drive is applied to the SCR gate and the SCR and the alarm bell are both off, but the op-amp goes to positive saturation at 'high' temperatures and thus drives the SCR and the alarm bell on. The SCR specified in the circuit has a mean current rating of only 2 A, so the alarm bell (a self-interrupting type) must be selected with this point in mind.

The two temperature-alarm circuits that we have looked at so far are designed to use thermistors with nominal resistances of 5 kΩ as their temperature-sensing elements, and these thermistors dissipate several milliwatts of power under working conditions. In some special applications this power dissipation may cause enough self-heating of the thermistor to upset the thermal sensing capability of the device. In such cases an alternative type of temperature-sensing device may have to be used.

Ordinary silicon diodes have temperature-dependent forward volt-drop characteristics, and can thus be used as temperature-sensing elements. Typically, a silicon diode gives a forward volt drop of about 600 mV at a current of 1 mA. If this current is held constant, the volt drop changes by about –2 mV for each degree centigrade increase in diode temperature. All silicon diodes have similar thermal characteristics. Since the power dissipation of the diode is a mere 0.6 mW under the above condition, negligible self-heating takes place in the device, which can thus be used as an accurate temperature sensor.

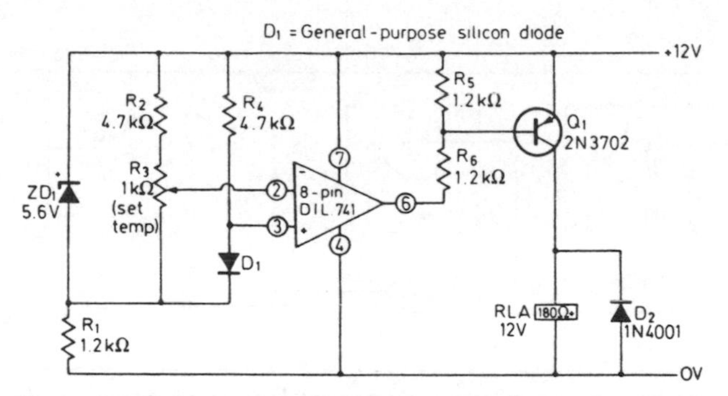

Figure 3.9 Relay-output over-temperature alarm using silicon-diode temperature sensing element

Figure 3.9 shows how a general-purpose silicon diode can be used as a thermal sensing element in an op-amp over-temperature alarm circuit. Here, zener diode ZD_1 is wired in series with R_1 so that a constant 5.6 V is developed across the two potential dividers formed by R_2–R_3 and R_4–D_1. A virtually constant current thus flows in each of these dividers. A constant reference voltage is thus developed between the R_1–R_3 junction and pin 2 of the op-amp, and a temperature-dependent voltage with a coefficient of –2 mV/°C is developed between the R_1–R_3 junction and pin 3 of the op-amp. Thus a differential voltage with a coefficient of –2 mV/°C appears between pins 2 and 3 of the op-amp.

In practice this circuit is set up by simply raising the temperature of D_1 to the required over-temperature trip level, and then slowly adjusting R_3 so that the relay just turns on. Under this condition a differential temperature of about 1 mV appears between pins 2 and 3 of the op-amp, the pin 3 voltage being below that of pin 2, and Q_1 and the relay are driven on. When the temperature falls below the trip level, the pin 3 voltage rises above that of pin 2 by about 2 mV/°C change in temperature, so Q_1 and the relay turn off. The circuit has a typical sensitivity of about 0.5°C, and can be used as an over-temperature alarm at temperatures ranging from sub-zero to above the boiling point of water.

It should be noted that the operation of the circuit can be reversed, so that it works as an under-temperature alarm, by simply transposing the pin 2 and pin 3 connections of the op-amp.

Finally in this section, *Figures 3.10* and *3.11* show the circuits of a pair of over-temperature alarms that give alarm outputs directly into loudspeakers. The *Figure 3.10* circuit generates a pulsed-tone alarm

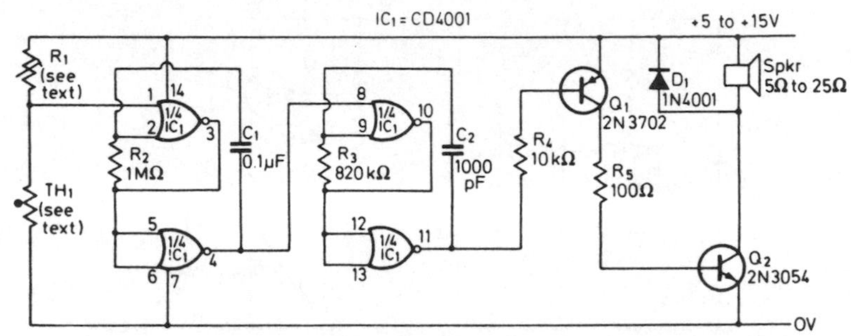

Figure 3.10 800 Hz pulsed-output non-latching over-temperature alarm

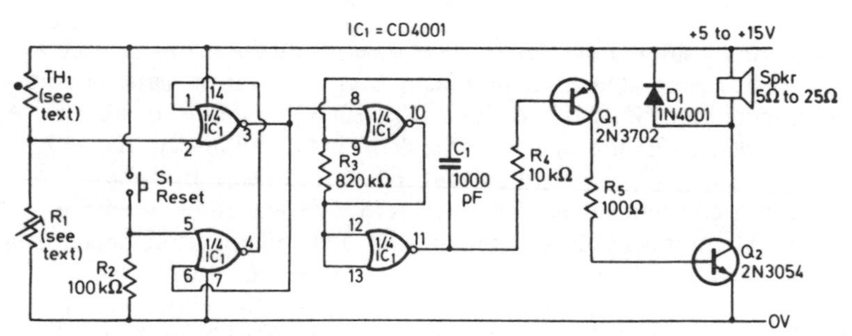

Figure 3.11 800 Hz monotone self-latching over-temperature alarm

signal, and gives non-latching operation. The *Figure 3.11* circuit generates an 800 Hz monotone alarm signal and gives self-latching operation. Both circuits are designed around a CD4001 COS/MOS IC.

The *Figure 3.10* and *3.11* circuits are identical to the *Figure 3.5* and *3.6* fire-alarm circuits respectively, except that their input activating signals are taken from the junction of the R_1-TH_1 potential divider rather than from the contacts of the thermostats.

An inherent feature of the *Figure 3.10* and *3.11* COS/MOS circuits is that they become enabled or disabled when their input activating voltages rise above or fall below a precisely defined 'threshold' value. This threshold voltage is not a fixed value, but is equal to a fixed percentage of the circuit's supply voltage, as in the case of a resistive potential divider. Consequently, these circuits switch from a disabled to an enabled state, or vice versa, when the R_1-TH_1 ratios go above or below a precisely defined value. This ratio is independent of the supply voltage, but is dependent on the threshold value of the individual CD4001 IC that is used in each circuit. The ratio has a nominal value of 50:50, but in practice may vary from 30:70 to 70:30 between individual ICs.

What the above paragraph means in practice is that the *Figure 3.10* and *3.11* circuits each turn on when their temperatures exceed a value that is preset by R_1. The circuits have typical sensitivities of about 0.5°C.

The basic *Figure 3.10* and *3.11* circuits can be used with any supply voltages in the range 5–15 V, and with speaker impedances in the range 5–25 Ω. The circuits give output powers in the range 0.25 W to 11.25 W, depending on the values of impedance and voltage that are used.

Under-temperature alarm circuits

The over-temperature alarm circuits of *Figures 3.7* to *3.11* can all be converted to give under-temperature alarm operation by making very simple alterations to their input connections, as shown in *Figures 3.12* to *3.16*.

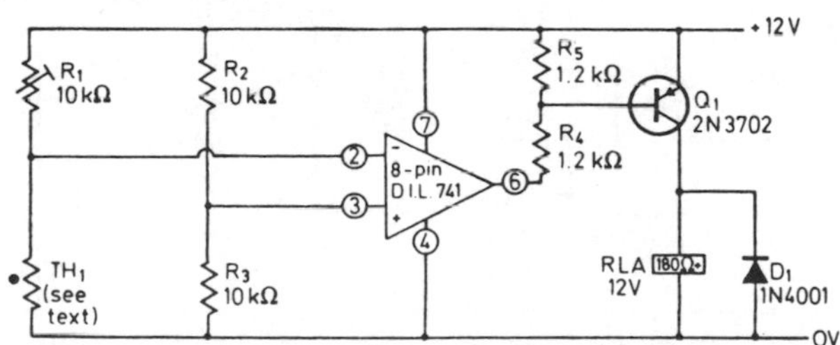

Figure 3.12 Relay-output precision under-temperature alarm

Figures 3.12 to *3.14* show how the circuits of *Figures 3.7* to *3.9* can be converted to under-temperature alarm operation by simply transposing the connections of pins 2 and 3 of their op-amps. *Figures 3.15* and *3.16* show how the circuits of *Figures 3.10* and *3.11* can be converted to under-temperature alarms by simply transposing their R_1 and TH_1 positions.

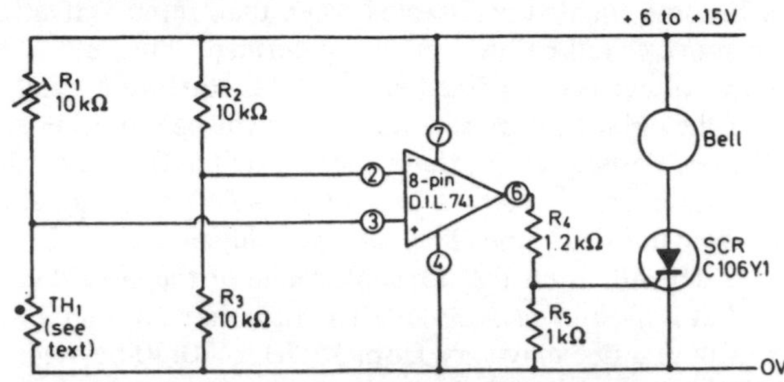

Figure 3.13 Direct-output precision under-temperature alarm

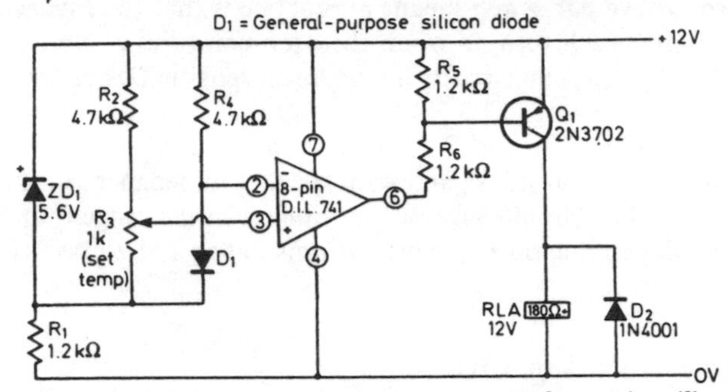

Figure 3.14 Relay-output under-temperature alarm using silicon-diode temperature sensing element

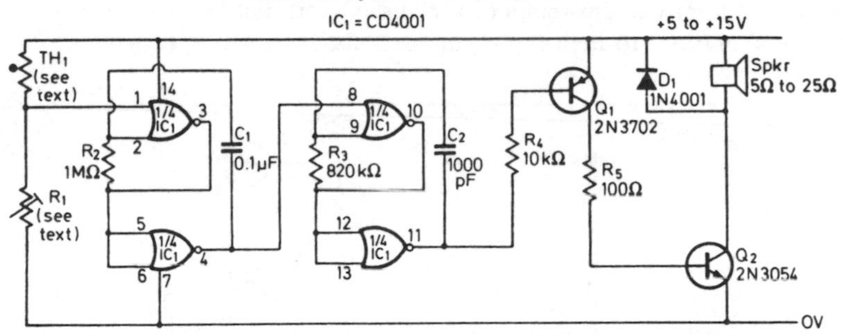

Figure 3.15 800 Hz pulsed-output non-latching under-temperature alarm

As a point of general interest, it may be noted that temperature alarms are normally used in the non-latching mode, so that the alarms are always off when the monitored temperature is within its preset limits. All the circuits of *Figures 3.7* to *3.15*, except *Figure 3.11*, are designed to give this kind of operation. If required, the circuits of *Figures 3.7, 3.9, 3.12* and *3.14* can each be made self-latching by wiring a spare set of n.o. relay contacts between the emitter and

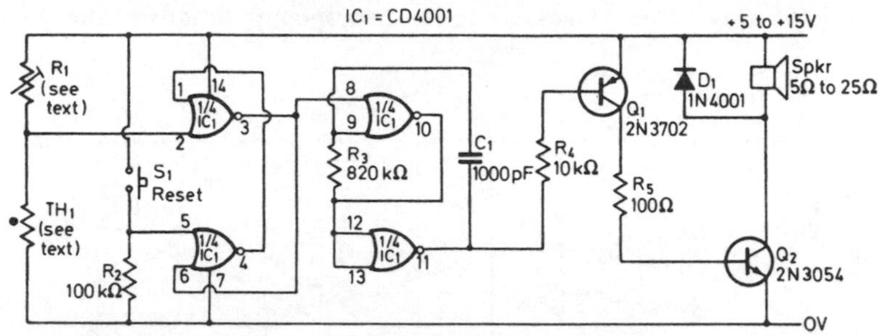

Figure 3.16 800 Hz monotone self-latching under-temperature alarm

collector of Q_1, and the circuits of *Figures 3.8* and *3.13* can each be made self-latching by wiring a 470 Ω resistor in parallel with the alarm bell.

Miscellaneous temperature alarms

The temperature alarm circuits that we have looked at so far are all designed to indicate an alarm condition when the temperature either goes above or below a preset level. In this final section of the chapter we show three other types of temperature-alarm system. Two of these systems are designed to indicate an alarm condition if the temperature deviates from a preset level by more than a preset amount, and the third system gives an alarm indication if two monitored temperatures differ by more than a preset amount. All three alarm systems are designed to give a relay output, which can be used to operate any type of audible or visual alarm device.

Figures 3.17 and *3.18* show the circuits of a pair of temperature-deviation alarms, which give an alarm indication if the temperature deviates from a preset level by more than a preset amount. The *Figure 3.17* circuit has independent over-temperature and under-temperature relay outputs, while the *Figure 3.18* circuit has a single relay output that activates if the temperature goes above or below preset levels.

Both circuits are made by combining the basic over-temperature and under-temperature circuits of *Figures 3.7* and *3.12.* The right (over-temperature) half of each circuit is based on that of *Figure 3.7,* and the left (under-temperature) half is based on that of *Figure 3.12.* Both halves of the circuit share a common $R_1 - TH_1$ temperature-sensing network, but the under-temperature and over-temperature switching levels of the circuits are independently adjustable. Each of the two op-amp outputs of the *Figure 3.17* circuit is taken to independent transistor-relay output stages, while the two op-amp outputs of the

R1 Set half-supply volts
R2 Set under-temp trip
R3 Set over-temp trip

Figure 3.17 Temperature-deviation alarm with independent over/under-temperature relay outputs

Figure 3.18 circuit are taken to a single transistor-relay output stage via the $D_1 - D_2$ gate network. The procedure for setting up the two circuits is as follows.

First set R_2 and R_3 to roughly mid-travel; then, with the thermistor at its normal or mid-band temperature, adjust R_1 so that half-supply

R1 Set half-supply volts
R2 Set under-temp trip
R3 Set over-temp trip

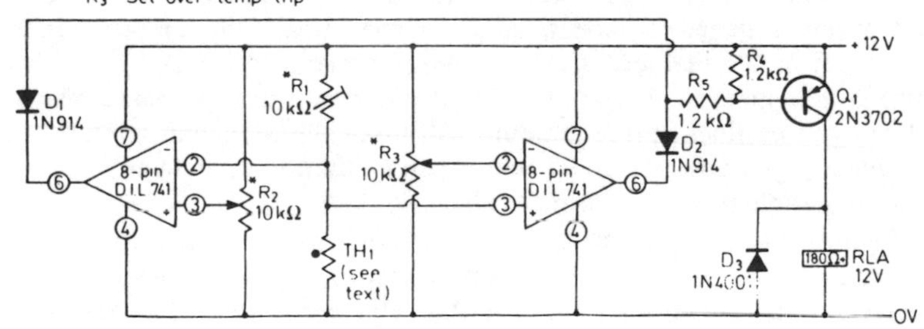

Figure 3.18 Temperature-deviation alarm with single relay output

volts are developed across TH_1. Now fully rotate the R_2 slider towards the positive supply line, rotate the R_3 slider towards the zero volts line, and check that no alarm condition is indicated (relays off). Next, reduce the TH_1 temperature to the required under-temperature trip level, and adjust R_2 so that the appropriate relay goes on to indicate an alarm condition. Now increase the temperature slightly, and check that the relay goes off. Finally, increase the temperature to the required over-temperature trip level, and adjust R_3 so that the appropriate relay goes on to indicate the alarm condition. All adjustments are then complete, and the circuits are ready for use.

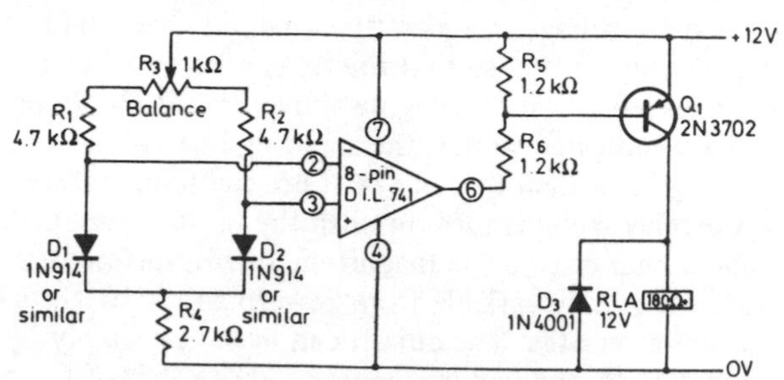

Figure 3.19 Differential-temperature alarm, relay output

Finally, to complete this chapter, *Figure 3.19* shows how a pair of silicon diodes can be used as temperature-sensing elements in a differential-temperature switch, which turns on only when the temperature of D_1 is more than a preset amount greater than that of D_2, and is not influenced by the absolute temperature of the two diodes. Circuit operation is as follows.

D_1 and D_2 are general-purpose silicon diodes, and are used as temperature-sensing elements. A standing current can be passed through D_1 from the positive supply line via R_3-R_1 and R_4, and a similar current can be passed through D_2 via R_3-R_2 and R_4. The relative values of these currents can be adjusted over a limited range via R_3, thus enabling the forward volt drops of the diodes to be equalised, so that they give zero differential output when they are both at the same temperature.

Suppose then that the diode voltages have been equalised in this way, so that zero voltage differential exists between them. If now the temperatures of both diodes are raised by 10°C, the forward voltages of both diodes will fall by 20 mV, and zero voltage differential will still exist between them. The circuit is thus not influenced by identical changes in the temperatures of D_1 and D_2.

Suppose, on the other hand, that the temperature of D_2 falls 1°C below that of D_1. In this case the D_2 voltage will rise 2 mV above that of D_1, so the pin 3 terminal of the op-amp will go positive to the pin 2 terminal, and the op-amp will go into positive saturation and hold Q_1 and the relay off. Finally, suppose that the temperature of D_2 rises

1°C above that of D_1. In this case the D_2 voltage will fall 2 mV below that of D_1, so the op-amp will go into negative saturation and drive Q_1 and the relay on. Thus the relay turns on only when the temperature of D_2 is above that of D_1 (or when the temperature of D_1 is below that of D_2). The circuit has a typical sensitivity of about 0.5°C.

In the explanation above it has been assumed that R_3 is adjusted so that the D_1 and D_2 voltages are exactly equalised when the two diodes are at the same temperature, so that the relay goes on when the D_2 temperature rises a fraction of a degree above that of D_1. In practice, R_3 can readily be adjusted so that the standing bias voltage of D_2 is some millivolts greater than that of D_1 at normal temperatures, in which case the relay will not turn on until the temperature of D_2 rises some way above that of D_1. The magnitude of this differential temperature trip level is fully variable from zero to about 10°C via R_3, so the circuit is quite versatile. The circuit can be set up simply by raising the temperature of D_2 the required amount above that of D_1, and then carefully adjusting R_3 so that the relay just turns on under this condition.

CHAPTER 4

LIGHT-SENSITIVE ALARM CIRCUITS

Light-sensitive alarm systems have a number of important applications in the home and in industry. They can be made to activate when light enters a normally dark area, such as the inside of a storeroom or a wall safe, or they can be used to sound an alarm when an intruder or object enters a prohibited area and breaks a projected light-beam. They can be used as smoke-sensitive alarms, and alarms that activate when the light level goes above or below a preset level.

Light-sensitive alarms may be designed to give an audible loudspeaker output, an alarm-bell output, or a relay output that can be used to switch any kind of audio/visual warning device. A wide range of alarm types are described in this chapter; most use a cadmium sulphide LDR (light-dependent resistor) as a light-sensing element or photocell; this device acts as a high resistance (typically hundreds of kilohms) under dark conditions, and a low resistance (typically a few hundred ohms or less) when brightly lit. Note that all circuits shown here will work well with almost any general-purpose LDRs with face diameters in the range 3 mm to 12 mm; no precise LDR types are thus specified in these circuits, although notes on their selection are given where applicable. Finally, note that this chapter deals only with 'visible-light' sensing circuits; infra-red (IR) 'invisible-light' light-beam alarms are dealt with in Chapter 8.

Simple light-sensitive alarm circuits

Figures 4.1 and *4.2* show a couple of very simple light-sensitive alarms, which turn on and self-latch when the illumination changes from near-dark to a moderately high level. These alarms are intended to sound when light enters a normally dark area, such as the inside of a storeroom or a wall safe.

In the *Figure 4.1* circuit, an alarm bell is wired in series with an SCR, which is wired in the self-latching mode via R_3 and S_1. The gate drive of the SCR is derived from the R_1–LDR–R_2 potential divider. Under dark conditions the LDR resistance is very high, so negligible voltage appears across R_2, and the SCR and the alarm are both off.

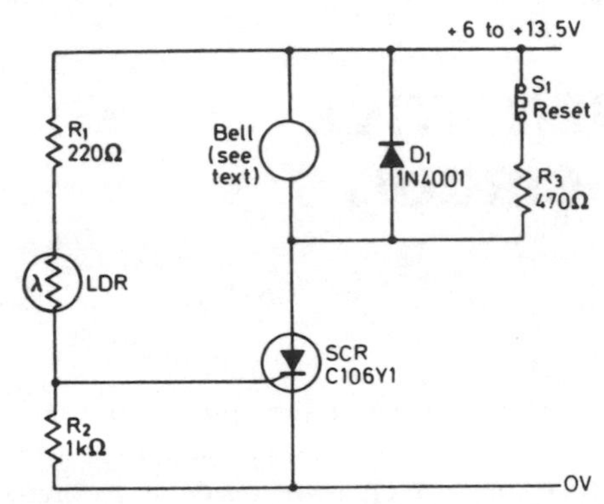

Figure 4.1 Simple light-activated alarm, bell output

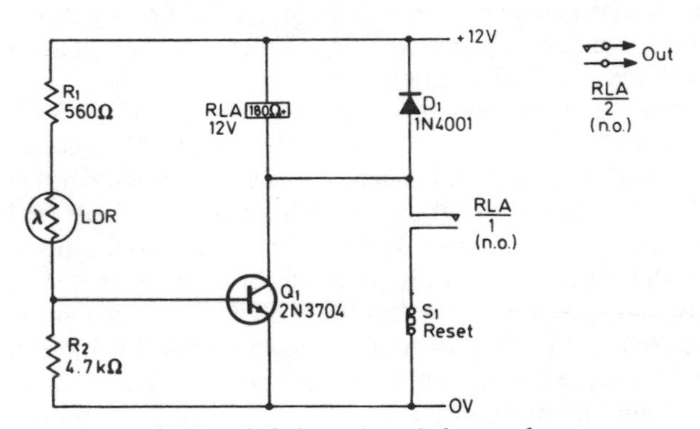

Figure 4.2 Simple light-activated alarm, relay output

When light enters the normally dark area and falls on the LDR face, the LDR resistance falls to a fairly low value; if this value is less than 10 kΩ or so, enough current flows through R_2 to turn the SCR and the alarm on, and the circuit then self-latches into the 'on' mode via R_3. Most LDRs give a resistance of less than 10 kΩ when exposed to low-intensity room lighting or to the light of a torch, so this circuit operates as soon as it is exposed to a moderate degree of illumination.

The *Figure 4.1* circuit can be used with any low-voltage self-interrupting alarm bell that draws a current less than 2 A. The circuit supply volts must be 1.5 V greater than the bell operating voltage. The design can be made non-latching by eliminating R_3 and S_1.

The *Figure 4.2* circuit operates in the same basic way as described above, except that an npn transistor is used in place of the SCR, and a relay is used in place of the alarm bell. The circuit can be used to activate any type of alarm device via the *RLA*/2 contacts. The relay

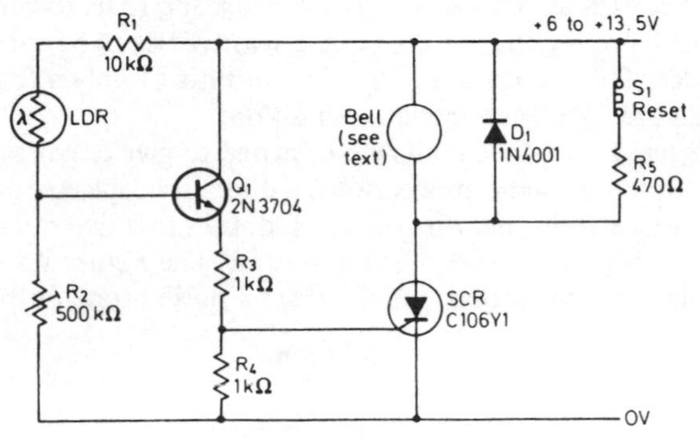

Figure 4.3 Improved light-activated alarm, bell output

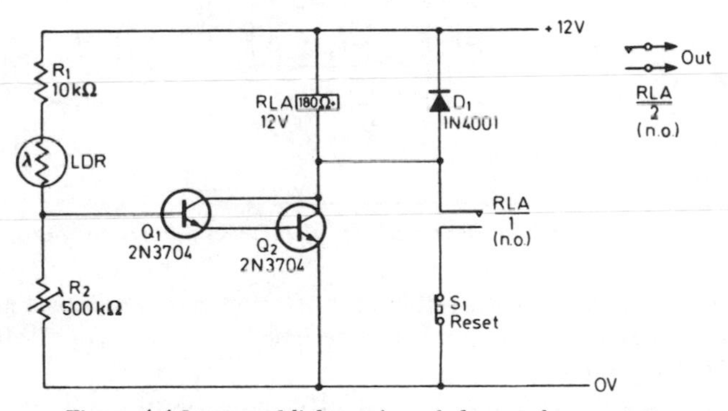

Figure 4.4 Improved light-activated alarm, relay output

can be any 12 V type with a coil resistance of 180 Ω or greater, and with two or more sets of n.o. contacts. The circuit can be made non-latching by eliminating S_1 and the *RLA*/1 contact connections.

The sensitivities of the *Figure 4.1* and *4.2* circuits can easily be increased so that the alarms turn on when only a very small amount of light falls on the LDR face. *Figures 4.3* and *4.4* show how.

The sensitivity of the *Figure 4.1* circuit is increased by replacing R_2 with a high-value variable resistor, and by interposing an emitter-follower buffer stage between the LDR potential divider and the gate of the SCR, as shown in *Figure 4.3.* The sensitivity of the *Figure 4.2* circuit is increased by replacing R_2 with a high-value variable resistor, and by replacing Q_1 with a super-alpha-connected pair of transistors, as shown in *Figure 4.4.*

The *Figure 4.3* and *4.4* circuits can both be turned on by LDR resistances as high as 200 kΩ, i.e. by exposing the LDR to very small amounts of light. R_2 enables the circuit sensitivities to be varied over a wide range. The circuits draw standby currents of only a few micro-amps when the LDR is under dark conditions.

The *Figure 4.1* to *4.4* circuits are designed to give either alarm-bell or relay outputs. In some applications, a direct loudspeaker output may be more suitable, and *Figures 4.5* and *4.6* show two circuits of this type; both give a low-level 'siren' output. The *Figure 4.5* circuit gives non-latching operation, and develops a pulsed-tone output signal.

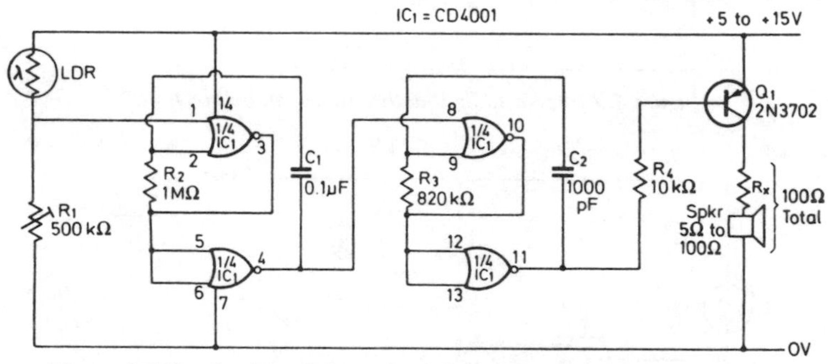

Figure 4.5 Non-latching light-activated alarm gives pulsed-tone output

The *Figure 4.6* circuit gives self-latch operation, and develops a mono-tone output signal. Both circuits are designed around a type CD4001 COS/MOS digital IC.

In the *Figure 4.5* circuit, the two left-hand gates of the IC are wired as a low-frequency (6 Hz) gated astable multivibrator that is activated by the light level, and the two right-hand gates are wired as an 800 Hz astable multivibrator that is activated via the 6 Hz astable. Under dark conditions both multivibrators are inoperative, and the circuit consumes a low standby current. Under bright conditions, on the other hand, both astables are activated, and the low frequency circuit pulses the 800 Hz astable on and off at a rate of 6 Hz, so a pulsed 800 Hz tone is generated in the speaker.

In the self-latching circuit of *Figure 4.6,* the two left-hand gates of the IC are wired as a simple bistable multivibrator, and the two right-hand gates are wired as a gated 800 Hz astable multivibrator that is activated via the bistable. Under dark conditions the output of the

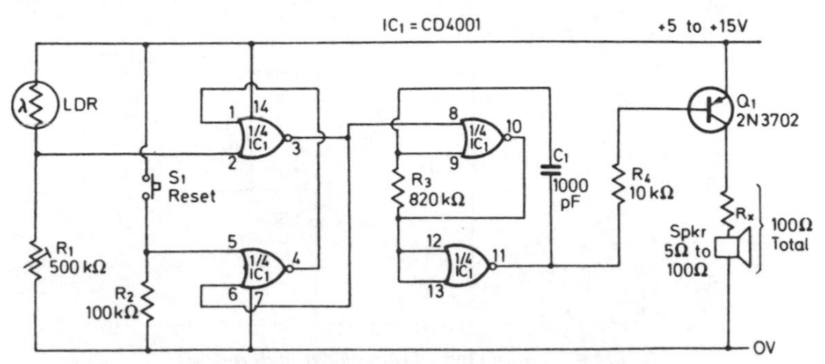

Figure 4.6 Self-latching light-activated alarm gives monotone output

bistable is normally high, and the astable is disabled, so the circuit consumes only a small standby current. When the LDR is exposed to bright light the bistable changes state and its output locks into the low state and activates the astable multivibrator. An 800 Hz tone signal is generated in the speaker under this condition. Once it has been activated, the circuit can only be turned off again by removing the illumination from the LDR and briefly closing 'reset' switch S_1, at which point the output of the bistable resets to the high state.

Note that the *Figure 4.5* and *4.6* circuits give output powers of only a few milliwatts, but that these levels can be boosted to as high as 18 W by replacing their Q_1 output stages with the power-boosting circuits of *Figures 1.9* or *1.10.*

Light-beam alarm circuits

Alarm circuits of this type are intended to activate when a person or object enters or 'breaks' a projected light-beam. *Figures 4.7* and *4.8* show two very simple 'visible-light' light-beam alarm circuits.

In the *Figure 4.7* circuit, the SCR is wired in the self-latching mode, uses a self-interrupting bell as its anode load, and has its gate current taken from the potential divider formed by R_1 and the LDR. Normally, the LDR is brightly illuminated via a light-beam formed by a remotely placed lamp and lens system, so the LDR acts as a low resistance under this condition, and insufficient voltage is developed at the R_1-LDR junction to turn the SCR on. When a person or object enters the light beam, the beam is broken and the resistance of the LDR rises to a fairly high value. Under this condition enough voltage is developed at the R_1-LDR junction to turn the SCR on, so the alarm goes on and self-latches.

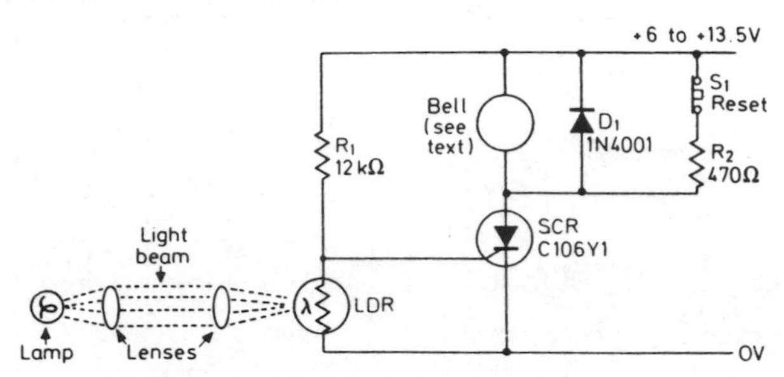

Figure 4.7 Simple light-beam alarm, bell output

The *Figure 4.8* circuit operates in the same basic way as described above, except that an npn transistor is used in place of the SCR, and a relay is used in place of the alarm bell. The circuit can be used to activate any type of alarm device via the *RLA*/2 contacts. The relay can be any 12 V type with a coil resistance of 180 Ω or greater, and with two or more sets of n.o. contacts. The circuit can be made non-latching by eliminating S_1 and the *RLA*/1 contact connections.

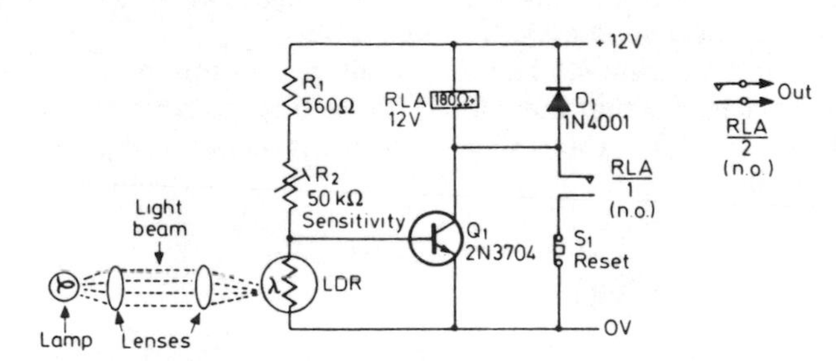

Figure 4.8 Simple light-beam alarm, relay output

The LDRs used in the *Figure 4.7* and *4.8* circuits can be any types that offer resistances of less than 1 kΩ under the illuminated condition, and more than 3 kΩ under the interrupted condition. The sensitivities of these circuits can readily be increased, so that the circuits can be used with any type of LDR, by using the connections shown in *Figures 4.9* and *4.10* respectively.

The *Figure 4.7* circuit is modified by interposing an emitter-follower stage between the R_1-LDR junction and the gate of the SCR, and by using a high-value variable resistor in the R_1 position, as shown in *Figure 4.9.* The *Figure 4.8* circuit is modified by using a super-alpha pair of transistors in place of Q_1, and by increasing the values of the potential divider components, as shown in *Figure 4.10.*

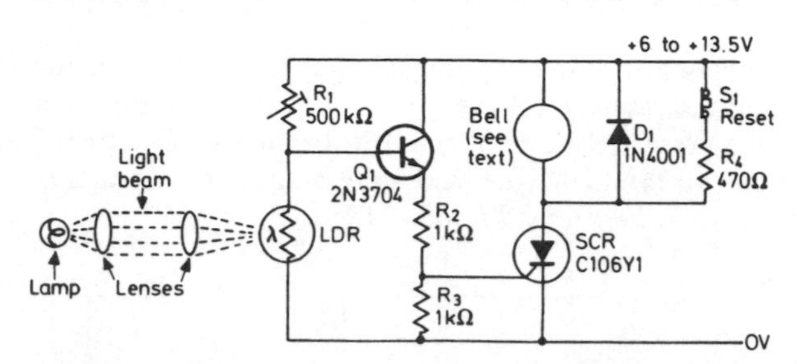

Figure 4.9 Improved light-beam alarm, bell output

The circuits of *Figures 4.7* to *4.10* can perform as useful intrusion detectors. They are inexpensive and easy to build. The lamps that activate them can be powered from either a.c. or d.c. supplies. A disadvantage of each circuit, however, is that it can be disabled by directing a bright light on to the LDR face. If this light has an intensity greater than that of the normal light beam, an intruder can walk through the beam without activating the alarm. This vulnerability of the basic light-beam alarm can be overcome in a number of ways.

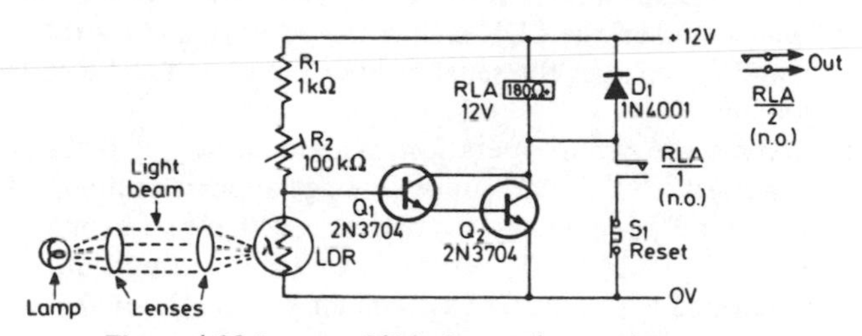

Figure 4.10 Improved light-beam alarm, relay output

The reader should note at this point that 'visible-light' beam systems are actually suitable for use only in low security applications such as object counting and giving automatic operation of electric doors and lights, etc. They are not suitable for use in high security applications such as burglar detection, because the beam is readily visible; for true high security, 'invisible light' infra-red beam systems (see Chapter 8) should be used.

Returning to LDR visible-light circuits, one system that beats the vulnerability problem uses a code-modulated light-beam, plus code-sensitive detector circuitry in its alarm section; the system cannot be disabled by shining a light on the LDR, since the alarm is sensitive only to the correct code signals. Such systems are fairly complex and expensive.

Another way of overcoming the vulnerability problem is to use an alarm circuit that activates if the intensity of the LDR illumination varies from a preset value, i.e. if the light-beam is broken or if a bright light is shone on the LDR face. A couple of practical circuits of this type are shown in *Figures 4.11* and *4.12*.

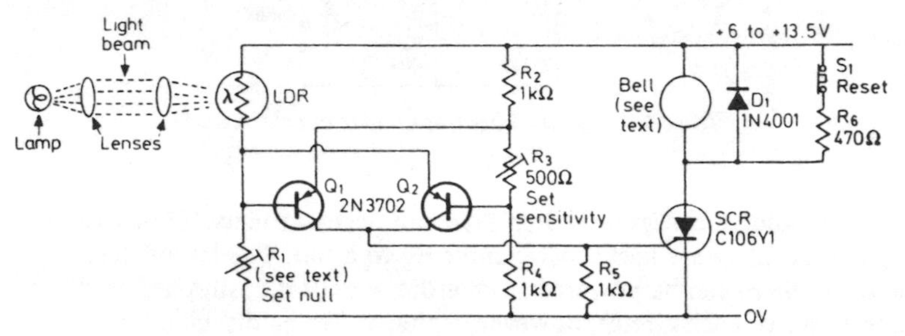

Figure 4.11 'Unbeatable' light-beam alarm, bell output

In the *Figure 4.11* circuit, the LDR is wired in a bridge circuit formed by $R_1-R_2-R_3-R_4$ and the LDR, and Q_1-Q_2 are used as a bridge-balance detector and SCR driver. R_1 is adjusted so that the bridge is balanced when the LDR is illuminated normally by the light-beam, and R_3 enables the sensitivity of the circuit to be varied over a reasonable range.

To understand the circuit operation, assume initially that R_3 is replaced by a short, so that half-supply voltage appears at the R_2-R_4 junction, and that R_1 is adjusted for balance, so that half-supply voltage appears at the *LDR*–R_1 junction. Under this condition, zero voltage is developed between the base and emitter of Q_1 or Q_2, so both transistors are cut off and zero current flows into the gate of the SCR. The alarm is thus off under this condition.

Suppose now that the light-beam is interrupted, so that the LDR resistance rises. Under this condition the voltage at the *LDR*–R_1 junction falls to a value lower than that on the R_2-R_4 junction, so a forward voltage appears between the base and emitter of Q_1. If this voltage exceeds 650 mV or so, Q_1 is driven on and its collector current feeds into the SCR gate, and the alarm circuit then turns on and self-latches.

Alternatively, suppose that the beam is not interrupted, but that a light with an intensity greater than that of the beam is shone on the LDR face. In this case the LDR resistance falls, so the voltage at the *LDR*–R_1 junction rises above that of the R_2–R_4 junction. A forward voltage, thus appears between the base and emitter of Q_2; if this voltage exceeds 650 mV or so, the transistor is driven on and its collector current feeds into the SCR gate, driving the alarm on. The alarm thus activates if the light intensity on the LDR face changes sufficiently to forward bias either transistor.

In the practical circuit of *Figure 4.11*, R_3 is wired in series with the R_2–R_4 potential divider, and enables a preset forward-bias voltage to be applied to the base–emitter junctions of Q_1 and Q_2, so that the circuit's sensitivity can be controlled. If, for example, a preset bias of 500 mV is applied to each transistor, the LDR only has to produce an additional change of 150 mV to turn one or other of the transistors on and thus activate the alarm. The circuit can thus be adjusted to a high degree of sensitivity, so that its immunity to 'disabling' by intruders can be as good as, or better than, that of even the most expensive modulated-light-beam systems.

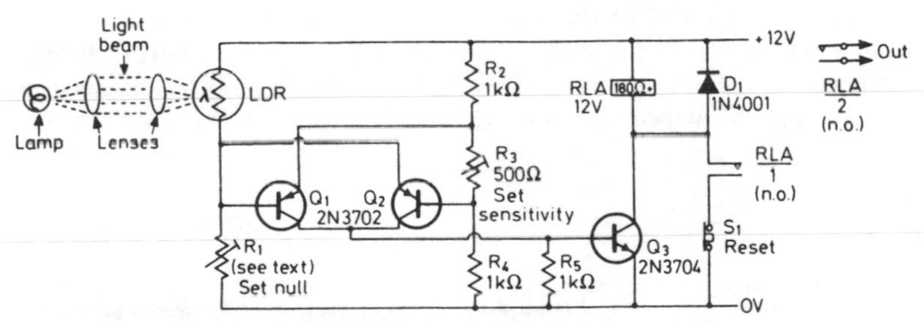

Figure 4.12 'Unbeatable' light-beam alarm, relay output

The *Figure 4.12* circuit is similar to that described above, except that the SCR and alarm bell are replaced by an npn transistor and a self-latching relay. This circuit can be used to activate any type of electrical alarm device via the relay contacts.

The LDR used in these circuits can be any type with a resistance in the range 200 Ω to 2 kΩ when illuminated by the light beam. R_1 should have a maximum value roughly double that of the LDR under the above condition, so that the two resistances are roughly equal when R_1 is set at mid-value. The sensitivity of each circuit varies slightly with changes in supply voltage, and is greatest at higher voltage levels. If the circuits are to be used at very high sensitivities, therefore, the supply voltages should be stabilised. To set up the circuits for use, proceed as follows.

First, adjust R_1 so that half-supply voltage is developed at the $LDR-R_1$ junction when the LDR is illuminated via the light beam, and then adjust R_3 so that roughly 400 mV are developed across R_5. Now readjust R_1 to give a minimum reading across R_5; readjust R_3, if necessary, so that this reading does not fall below 200 mV. When the R_1 adjustment is complete, the bridge is correctly balanced. R_3 can then be adjusted to set the sensitivity to the required level. If R_3 is set so that zero voltage is developed across R_5, fairly large changes in light level will be needed to operate the alarm; if it is set so that a few hundred millivolts are developed across R_5, only small changes will be needed to operate the alarm.

Smoke-alarm circuits

Another useful type of light-sensitive alarm is the so-called 'smoke' alarm. This may be of either the 'reflection' type or the 'light-beam' type.

Figures 4.14 and *4.15* show two practical reflection-type smoke-alarm circuits. The heart of these systems is the smoke detector, shown in *Figure 4.13*. Here, a lamp and an LDR are placed next to one another in a light-excluding box, but are shielded from one another by a simple

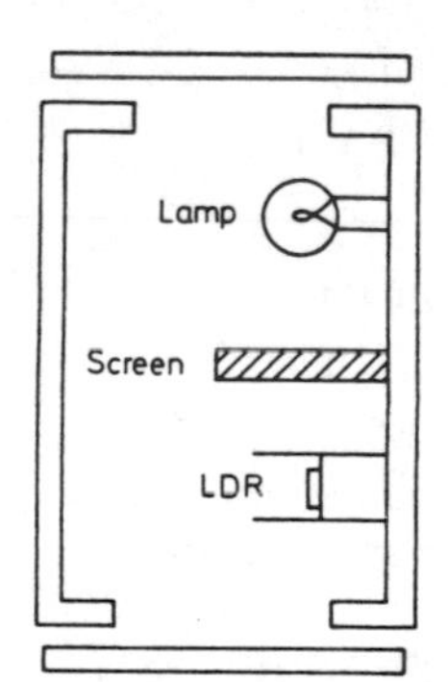

Figure 4.13 Sectional view of reflection-type smoke detector. Lamp provides light source, and heat to cause convection currents to draw air in from bottom of box and expel it through lid; inside of box is painted matt black; construction lets air pass through box but excludes external light

screen so that the light of the lamp does not fall directly on to the face of the LDR. The inside of the box is painted matt black, so that the LDR is not illuminated by light reflected from the inside of the box.

The construction of the box is such that its top and bottom are open to the air, but exclude external light. The lamp is placed near the top of the box, and inevitably generates a certain amount of heat. This heat rises out of the top of the box and sucks cooler air in through the bottom. Thus a continuous current of air is passed through the box

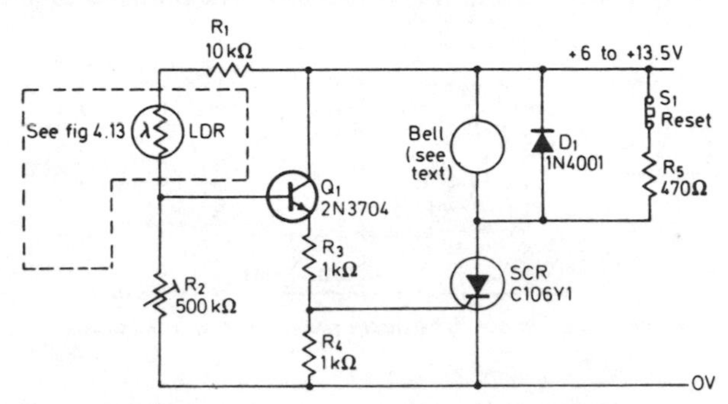

Figure 4.14 Reflection-type smoke alarm, bell output

and past the face of the LDR. If this air is smoke-free it is invisible and does not reflect the light of the lamp, so the LDR remains in darkness and presents a high resistance. Alternatively, if the circulating air is laden with smoke it reflects the light of the lamp on to the face of the LDR, and the LDR resistance falls to a fairly low value. This reduction of the LDR resistance can be used to operate a simple alarm circuit.

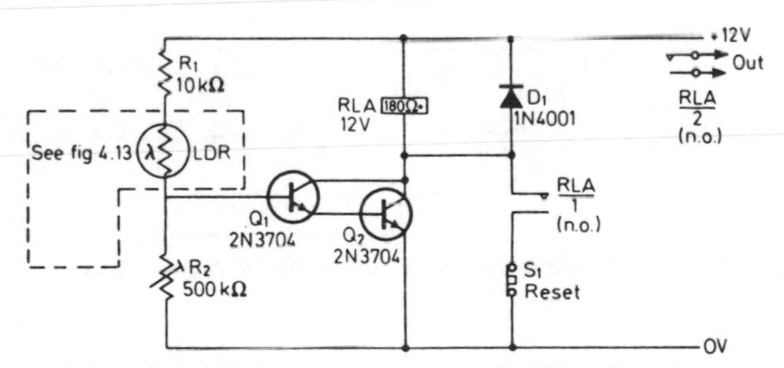

Figure 4.15 Reflection-type smoke alarm, relay output

Two suitable alarm circuits are shown in *Figures 4.14* and *4.15*; they are virtually identical to those shown in *Figures 4.3* and *4.4* respectively, and activate when the LDR resistance falls below a preset amount. When used in conjunction with the smoke detector of *Figure 4.13*, these circuits thus act as 'smoke' alarms. The circuits are simple and give reliable operation.

In the alternative 'light-beam' type of smoke alarm, a light-beam is projected across the protected area and on to the face of an LDR, which forms part of a sensitive detector circuit. When smoke enters the light-beam, the intensity of the illumination on the LDR face falls slightly, and the LDR resistance rises. This increase of resistance is used to activate the alarm, which thus responds to the presence of smoke.

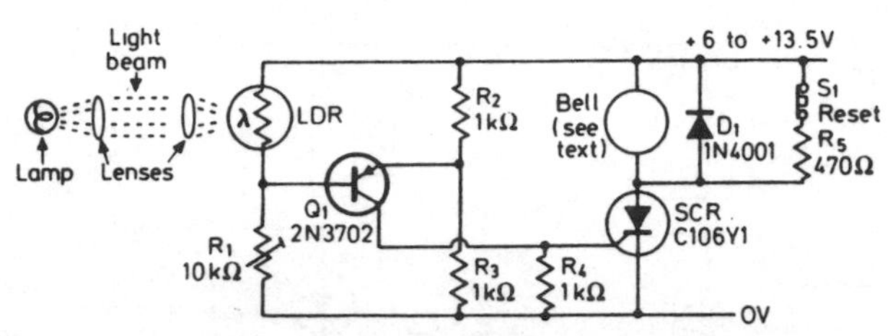

Figure 4.16 Simple light-beam smoke alarm, bell output

Two simple light-beam smoke-alarm circuits are shown in *Figures 4.16* and *4.17*. In the *Figure 4.16* circuit, the LDR and R_1–R_2–R_3 are wired in the form of a simple bridge. R_1 is adjusted so that the bridge is out of balance in such a way that Q_1 is not quite biased on when the LDR is normally illuminated by the light beam. Under this condition Q_1 passes negligible collector current into the gate of the SCR, so the alarm is off.

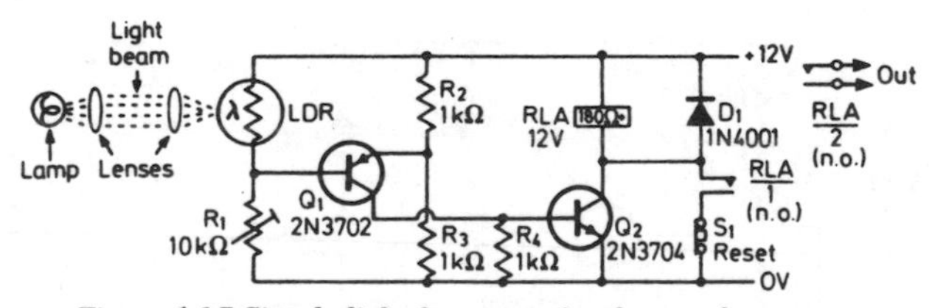

Figure 4.17 Simple light-beam smoke alarm, relay output

When smoke enters the light beam, the LDR resistance increases and throws the bridge out of balance in such a way that the base–emitter junction of Q_1 is appreciably forward biased. Under this condition, Q_1 passes substantial collector current into the gate of the SCR, and the alarm turns on and self-latches.

The *Figure 4.17* circuit is similar to that described above, except that an npn transistor and a self-latching relay are used in place of the SCR and alarm bell.

A snag with the *Figure 4.16* and *4.17* circuits is that their trigger points are slightly affected by variations in temperature, since the V_{be} characteristics of Q_1 are temperature dependent. The circuits are thus not suitable for use in conditions of large temperature variation or at high sensitivity levels.

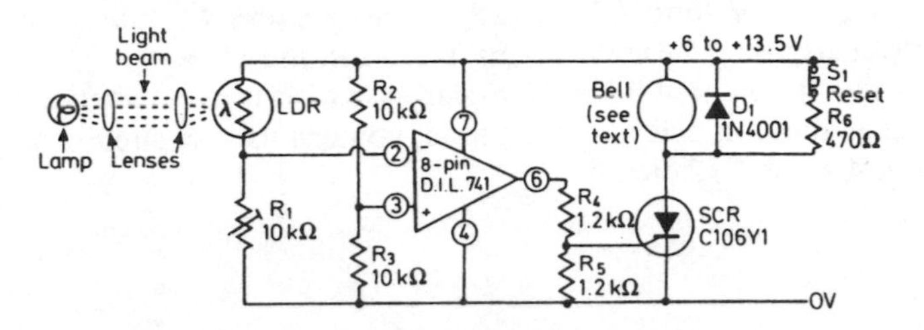

Figure 4.18 Sensitive light-beam smoke alarm, bell output

These snags are largely overcome in the two sensitive smoke-operated alarm circuits of *Figures 4.18* and *4.19.* Here, the LDR is again connected in the bridge network formed by the LDR and R_1–R_2–R_3, but in this case the operational amplifier is used as the bridge-balance detector. An outstanding feature of the op-amp is that its operating points are not greatly affected by variations in ambient temperature or supply voltage. Consequently these circuits give very stable operation.

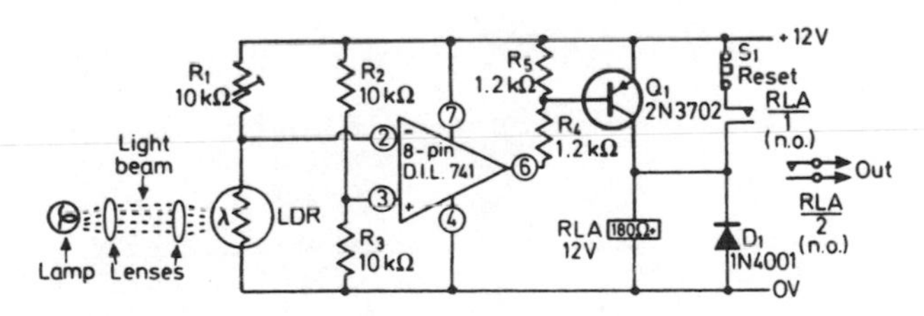

Figure 4.19 Sensitive light-beam smoke alarm, relay output

In the *Figure 4.18* circuit the op-amp is used to operate the alarm bell via the SCR, while in the *Figure 4.19* circuit it is used to operate the self-latching relay via Q_1. Note that if these circuits are to be used at their maximum sensitivity levels it may be necessary to feed the light-beam energising lamps from regulated power sources, so that the illumination levels of the light beams are stabilised to a reasonable degree.

'Dark' and 'light/dark' alarm circuits

Most of the circuits shown in this chapter can be used, or adapted for use, in applications other than those for which they were specifically designed. If you need 'dark-operated' circuits, which turn on when the light level falls below a preset value, you can use the circuits of *Figures 4.7* to *4.10* and *4.16* to *4.19* directly, or use *Figures 4.5* and *4.6* by simply transposing the connections of R_1 and the LDR.

Finally, if you need 'light/dark' alarms, which turn on when the light goes above or below a preset level, you can use the circuits of *Figures 4.11* or *4.12* directly.

CHAPTER 5

MISCELLANEOUS ALARM CIRCUITS

In each of the four preceding chapters we have looked at a specific class of alarm system. In this chapter we look at a miscellaneous range of alarm circuits that can also be used in the home or in industry.

The circuits presented here include liquid and steam-activated alarms, power-failure alarms, an ultrasonic beam alarm, and sound or vibration alarms.

Liquid and steam-activated alarms

Liquid- and steam-activated alarms have a number of uses. Liquid-activated alarms can be made to sound when the water in a bath or the liquid in a tank reaches a preset level, or when rain falls across a pair of contacts, or when flooding occurs in a cellar or basement, or when an impact wave is generated as a person or object falls into a swimming pool or tank.

Steam-activated alarms can be made to sound when high-pressure steam escapes from a valve or a fractured pipe, or when steam emerges from the spout of a kettle or container as the liquid reaches its boiling point.

Five liquid- or steam-activated alarm circuits are described. All the circuits use the same basic principle of operation. In each case, a pair of metal probes detects the presence or absence of the liquid or steam. In the absence of the medium the probes 'see' a near-infinite resistance, but in the presence of the medium the probe resistance falls to a relatively low value. This fall in resistance is detected and made to activate the alarm device. The resistance appearing across the probes under the alarm condition depends on the type of medium that is being detected. In the case of rain or tap water, the resistance may be less than a few kilohms, but in the case of steam or oil the resistance may be greater than several megohms.

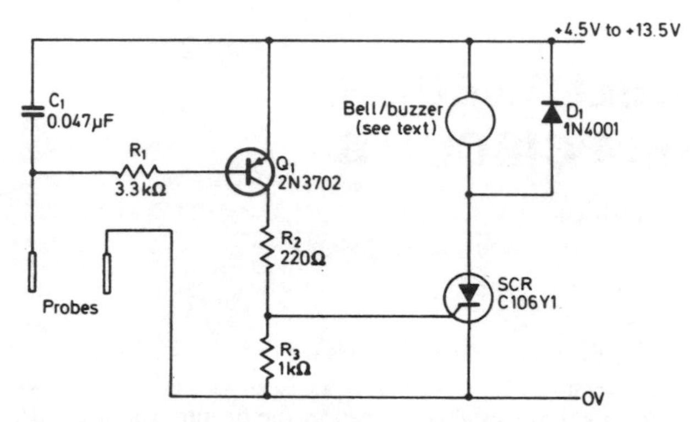

Figure 5.1 Simple liquid- or steam-activated alarm, bell/buzzer output

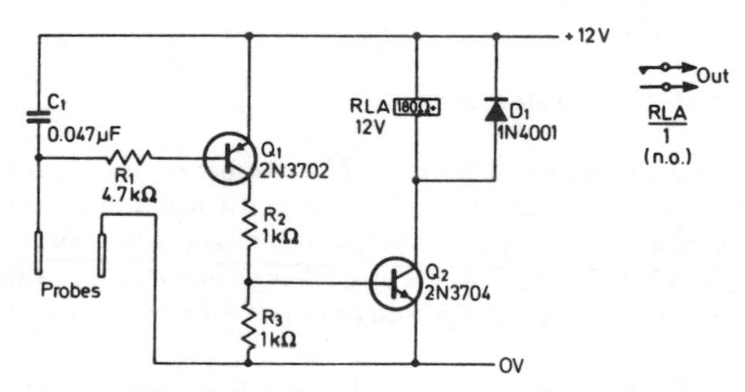

Figure 5.2 Simple liquid- or steam-activated alarm, relay output

Two simple liquid-activated alarm circuits, which can be activated by probe resistances up to about 500 kΩ, are shown in *Figures 5.1* and *5.2*. Both circuits operate in the same basic way, and give a non-latching form of operation. The former gives an alarm bell or buzzer output, and the latter gives a relay output that can be used to activate any type of alarm device via the relay contacts.

In the case of the *Figure 5.1* circuit, Q_1 is cut off when the probes are open-circuit, so the SCR and alarm are also off. When a resistance less than 500 kΩ or so is connected across the probes, Q_1 is biased on to such a level that its collector current turns on the SCR, and the alarm-bell or buzzer activates. Note that this bell or buzzer must be a self-interrupting device, so that the alarm turns off when the resistance is removed from the probes. The circuit supply voltage must be 1.5 V greater than the bell or buzzer operating voltage, and the alarm device must pass a current of less than 2 A.

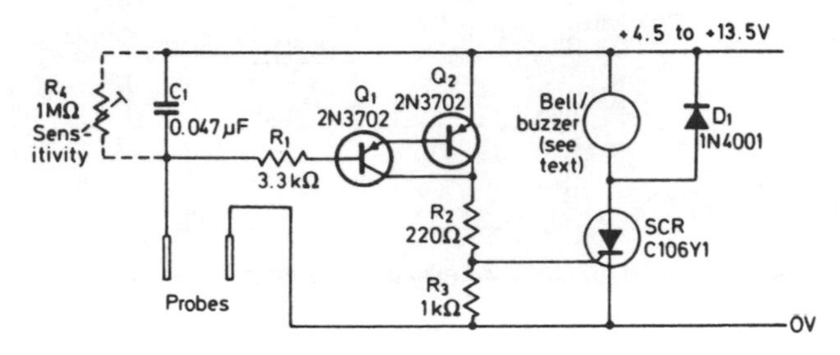

Figure 5.3 Sensitive liquid- or steam-activated alarm, bell/buzzer output

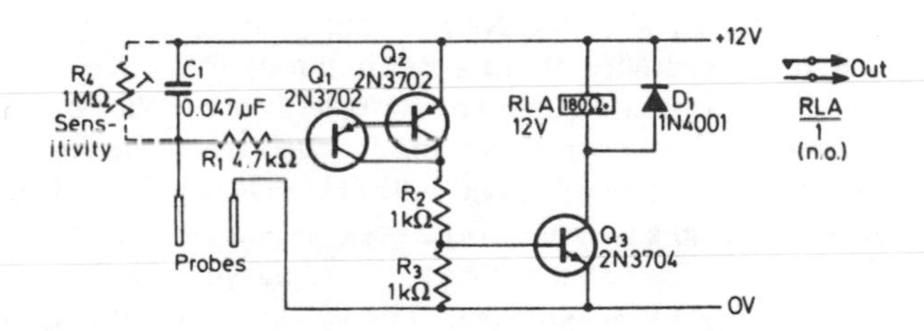

Figure 5.4 Sensitive liquid- or steam-activated alarm, relay output

The *Figure 5.2* circuit is similar to that described above, except that the SCR is replaced by an npn transistor, and a 12 V relay is used in place of the bell or buzzer. The relay must have a coil resistance more than 180 Ω. Its contacts can be used to activate any type of external alarm device.

The circuits of *Figures 5.1* and *5.2* can easily be modified so that they are activated by probe resistances up to about 20 MΩ. *Figures 5.3* and *5.4* show how. In each case, Q_1 is simply replaced by a super-alpha-connected pair of pnp transistors. These circuits also show how the sensitivity of the designs can be made variable by wiring a 1 MΩ preset resistor (shown dotted) across C_1. C_1 is used to protect the circuits against activation by spurious or radiated signals.

Finally, *Figure 5.5* shows the circuit of a liquid- or steam-activated alarm that gives a pulsed-tone low-level output signal directly into a

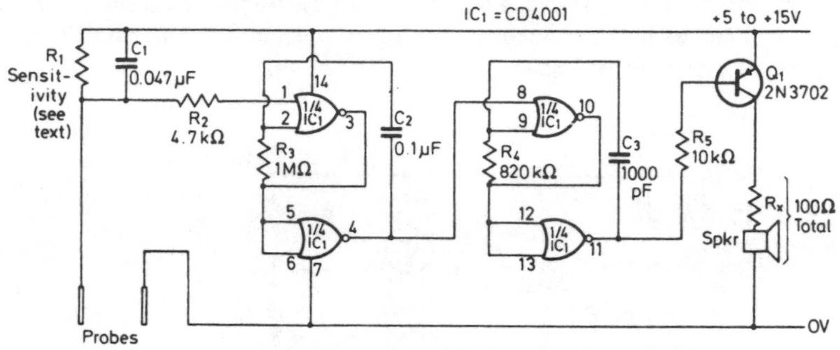

Figure 5.5 Liquid- or steam-activated alarm, pulsed-tone output

loudspeaker. The circuit uses a CD4001 COS/MOS IC as a gated pulse-tone generator, which feeds the speaker via Q_1. The generator is normally off, but turns on when the probe resistance falls below a value that is roughly equal to R_1. R_1 can thus be selected to give the circuit any desired sensitivity, up to a maximum of about 20 MΩ.

When this circuit is in the quiescent state, with the probes open-circuit, it consumes a total current of about 1 µA. When the circuit is activated it generates a tone of roughly 800 Hz that is pulsed on and off at a rate of 6 Hz, so a pulsed 800 Hz tone is generated in the speaker. The circuit is based on the *Figure 1.11* design and gives an output power of only a few milliwatts. The power can be boosted as high as 18 W by replacing the Q_1 output stage by one or other of the *Figure 1.9* and *1.10* power-boosting stages.

Power-failure alarm circuits

Electrical power-failure alarms can be made to activate when power is removed from a deep-freeze unit, or when a burglar cuts through power lines, or when a machine overloads and blows its fuses. Three useful power-failure alarm circuits are described in this section.

Figure 5.6 shows a simple relay-output power-failure alarm, which can be used to activate any type of external alarm device via the relay contacts. Here, the power-line input is applied to a step-down transformer, which gives an output of 12 V. This output is half-wave rectified by D_1 and smoothed by C_1, and the resulting d.c. is fed directly to the relay coil. The n.c. contacts of the relay are used to apply power to the external alarm device.

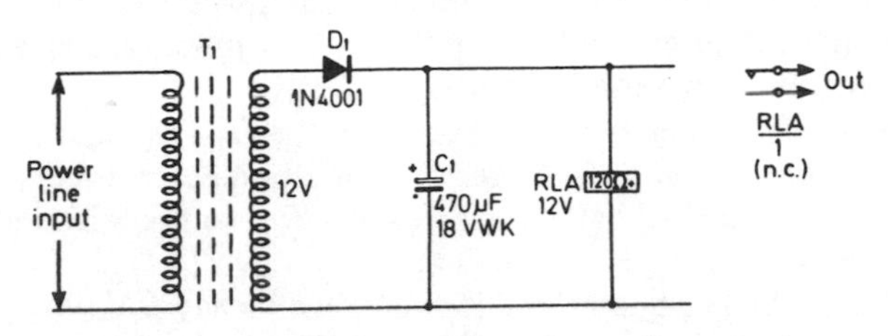

Figure 5.6 Simple power-failure alarm, relay output

Thus when power is applied to the circuit the relay is driven on, its contacts are open and the alarm is off, but when the power input is removed the relay turns off, so its contacts close and activate the alarm device.

The relay can be any 12 V type with a coil resistance of 120 Ω or greater, and with one or more sets of n.c. contacts. T_1 can be any power-line step-down transformer that gives a 12 V output at a current above 100 mA.

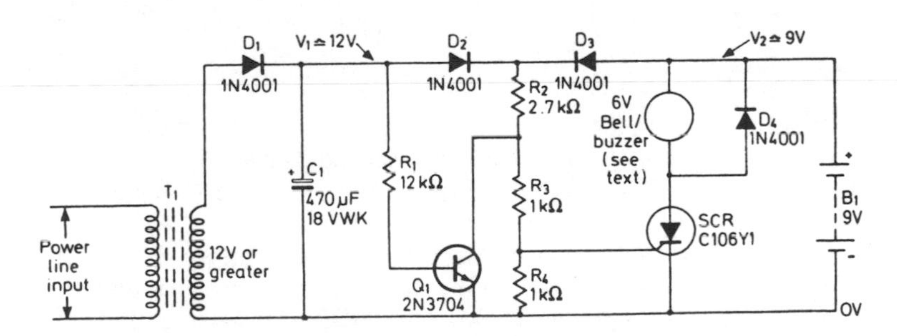

Figure 5.7 Power-failure alarm with bell/buzzer output

An alternative type of power-failure alarm is shown in *Figure 5.7*. Here, the power input is stepped down to 12 V by T_1 and is rectified and smoothed by D_1 and C_1, to give roughly 12 V d.c. at the $D_1 - D_2$

and D_2-D_3 junctions. The actual alarm device, which is a self-interrupting bell or buzzer, is used as the anode load of the SCR and is powered from a 9 V battery.

Normally, when power is applied, 12 V d.c. are developed at the D_1-D_2 and D_2-D_3 junctions, so Q_1 is driven to saturation via R_1, and the R_2-R_3 junction is pulled down to zero volts. Under this condition zero drive is applied to the SCR gate, so the alarm is off and D_3 is reverse-biased by the 12 V on the D_2-D_3 junction and no current is drawn from the 9 V battery.

When power is removed from the input, the D_1-D_2 junction falls to zero volts and Q_1 turns off. Under this condition current feeds to the SCR gate from the 9 V battery via $D_3-R_2-R_3$, so the SCR and the alarm turn on.

The alarm bell or buzzer in this circuit can be any 6 V self-interrupting type that draws less than 2 A. The step-down transformer can be any type that gives a 12 V output at a current of a few milliamps.

If this circuit is modified for use with alternative voltages, it is essential that the voltage at the D_1-D_2 junction be at least 2 V greater than the battery volts.

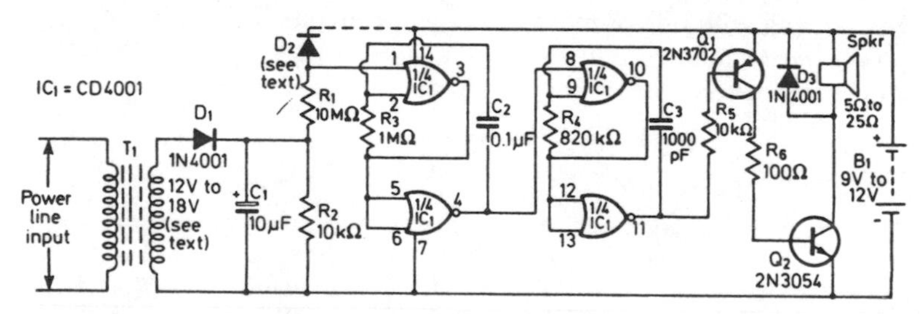

Figure 5.8 Power-failure alarm with pulsed-tone output

Finally, *Figure 5.8* shows the circuit of a power-failure alarm that gives a medium-power (about 10 W) pulsed-tone alarm signal directly into a loudspeaker. Here, a CD4001 COS/MOS IC is wired as a gated pulse-tone generator, of the type shown earlier in *Figures 1.11* and *5.5*, and feeds the speaker via a power-booster stage of the type shown in *Figure 1.9*.

In this circuit the power-line signal is again stepped down, rectified and smoothed by $T_1-D_1-C_1$. When power is applied the voltage across C_1 is greater than that of the supply battery, so the input gate of the IC is clamped to the battery positive-rail voltage via R_1 and the gate-input protection diode (shown dotted as D_2) of the IC, and the generator is gated off. Under this condition the circuit consumes only a small leakage current from the battery.

When power is removed from the input of the circuit the C_1 voltage falls to zero. Under this condition the IC is gated on, and an alarm signal is generated in the speaker. This signal has a basic frequency of 800 Hz, and is pulsed on and off at 6 Hz. The step-down transformer used can be any type that causes a voltage greater than that of the battery to be developed across C_1.

Proximity alarm circuits

As the name indicates, proximity alarms can be made to activate when a person or object touches or comes close to a sensing antenna or a conducting object attached to the antenna. Two practical proximity alarm circuits are described, both using the same basic principle of operation. One circuit (*Figure 5.9*) gives a relay output, the other (*Figure 5.10*) gives a direct alarm-bell or buzzer output.

Both circuits work on the capacitive loading principle, in which the gain of an r.f. oscillator is adjusted to a critical point at which oscillation is barely sustained, and in which the antenna forms part of the tank circuit. In these circuits one of the supply lines is grounded.

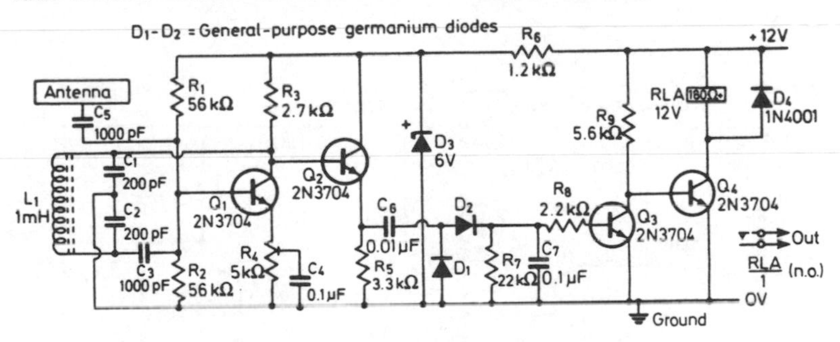

Figure 5.9 Relay-output proximity alarm

Consequently any increase in the antenna-to-ground capacitance, such as is caused by touching or nearing the antenna, causes enough damping of the tank circuit to bring the oscillator gain below the critical level, and the oscillator ceases to operate. This cessation of oscillation is then used to make the alarm generator activate.

Figure 5.9 shows the practical circuit of a relay-output system that uses the above principle. Here, transistor Q_1 is wired as a Colpitts oscillator, with gain adjustable via R_4, and the antenna is coupled to the base of Q_1 via C_5. The output of this oscillator, which operates at about 300 kHz, is made available at a low impedance level across R_5 via emitter-follower Q_2. This signal is rectified and smoothed via D_1–D_2–R_7 and C_7, to produce a positive bias that is fed to the base of Q_3 via R_8. Q_3 is wired as a common-emitter amplifier, with R_9 as a collector load, and Q_4 is wired as a common-emitter amplifier with the relay as its collector load and its base directly coupled to the collector of Q_3.

Thus when Q_1 is oscillating normally a positive bias is developed and drives Q_3 to saturation. Since Q_3 is saturated, its collector is at near-zero voltage, so zero bias is fed to the base of Q_4, and Q_4 and the relay are thus off under this condition.

When the antenna is touched or additionally loaded, the oscillator ceases to operate, so zero bias is developed by the rectifier-smoothing network and Q_3 is cut off. Since Q_3 is cut off, the base of Q_4 is taken directly to the positive supply line via R_9, so Q_4 and the relay are driven hard on under this condition.

The Q_1–Q_2 section of the circuit is fed from the regulated 6 V supply formed by R_6 and zener diode D_3, so oscillator stability is virtually independent of actual supply-line potential. Diode D_4 protects the circuit against damage due to the back e.m.f. from the relay as the circuit operates.

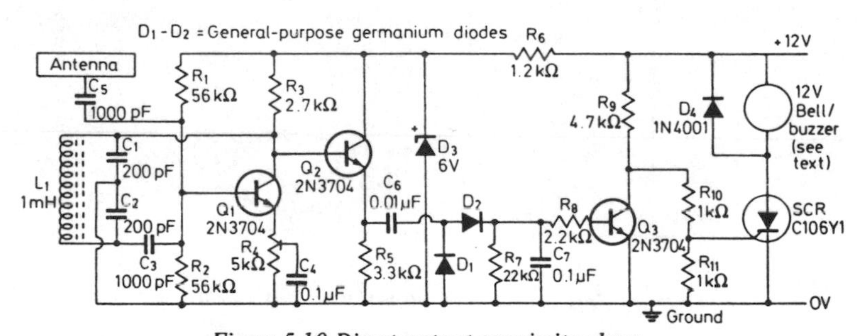

Figure 5.10 Direct-output proximity alarm

The direct-output version of the circuit, shown in *Figure 5.10*, is similar to that described above, except that an SCR is used in place of Q_4, and an alarm device is used directly as the anode load of the SCR.

This alarm device can be any self-interrupting bell or buzzer with a rating in the range 9–12 V at up to 2 A.

Note that the *Figure 5.9* and *5.10* circuits are both designed to give non-latch operation. The *Figure 5.9* circuit can be made self-latching by wiring a spare set of n.o. relay contacts between the collector and emitter of Q_4, and the *Figure 5.10* circuit can be made self-latching by wiring a 470 Ω resistor in parallel with D_4.

The two circuits are very simple to set up. First connect a suitable antenna, then turn R_4 towards the ground rail until the alarm just activates. Next, turn R_4 back a fraction so that the relay just turns off, then check that the alarm goes on when the antenna is touched or closely approached, and goes off again when the touch is removed. If necessary, adjust R_4 again for maximum sensitivity.

The final sensitivity of each circuit depends on the setting of R_4 and on the size of antenna used. If the antenna is very small, such as a short length of wire, the circuits will act as little more than touch alarms, but if the antenna is large, such as a sheet of metal, the circuits may be made sensitive enough to activate when a person approaches within a foot or two of the antenna. It pays to experiment with different types of antenna, to get the 'feel' of the circuits. Remember, however, that it is imperative that the antenna be well isolated from ground, and that one side of the circuit's power supply be taken to an effective ground connection. In some applications a floating artificial ground (such as a metal plate) can be used with advantage. If, for example, two metal plates are placed parallel a foot or so apart, and one is used as the antenna and the other as an artificial ground, the alarm will activate whenever a hand is placed between the two plates. Such a system can be used to sound an alarm if a hand is placed inside a small cabinet, etc.

Touch alarm circuits

Touch alarms are intended to activate when a person or object touches a fixed contact point. They work in a number of ways. They may be activated by touching and closing a simple microswitch, as in the contact alarm circuits described in Chapter 1, or they may work on the capacitive loading principle described in the preceding section of this chapter. Alternatively, they may be activated by the a.c. hum that is picked up by an electrical contact when it is touched by a human finger (in equipment that is connected to a.c. power lines), or by the relatively low resistance (less than a few megohms) that appears across a pair of contacts when they are bridged by a human finger. Three circuits of the latter two types are described in this section.

Figure 5.11 shows a simple but useful 'hum-detecting' touch alarm circuit. Here, one of the gates of a CD4001 COS/MOS IC is wired as a simple pulse-inverting amplifier, and has its high-impedance input terminal taken to a pick-up contact via R_2. The gate is effectively powered from a 5 V supply, derived from the 12 V line via R_3 and R_4, and is biased via R_1 so that its output is normally low.

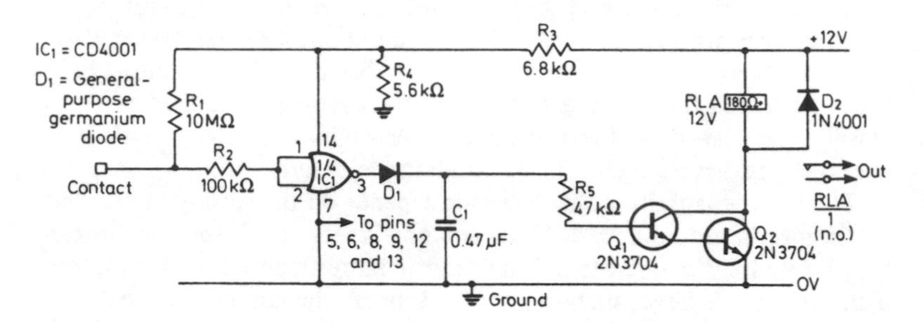

Figure 5.11 'Hum-detecting' touch alarm

When a pick-up signal with a peak amplitude greater than a couple of volts is applied to the pick-up contact, the output of the gate pulses on and off at line frequency, and a square wave with a peak amplitude of about 5 V is developed at the gate output terminal. This square wave is rectified and smoothed via D_1 and C_1, and the resulting d.c. is used to drive RLA on via Q_1–Q_2 and R_5.

Note when using this circuit that the low side of the 12 V supply must be grounded, and that any type of external alarm device can be activated via the circuit's relay contacts. The circuit draws a standby current of 1 mA. Since the circuit must be operated from a.c. power lines, this current drain should present no problems.

The pick-up contact of the above circuit should be limited in size to a few square centimetres. If the contact is to be placed more than about 10 cm away from the input terminal of the COS/MOS gate, the connecting leads must be screened to prevent the pick-up of unwanted signals.

A simple resistance-sensing touch alarm circuit is shown in *Figure 5.12.* Here, one of the gates of a CD4001 IC is again wired as a pulse inverter, but is powered directly from the 12 V supply line. The input of the inverter is strapped to the positive supply line via 10 MΩ resistor R_1, so its output is normally low. The output is used to drive Q_1 and the relay via R_3.

The circuit action is such that the output of the inverter is low and Q_1 and the relay are off when a resistance much greater than 10 MΩ appears across the touch contacts, but the output of the inverter goes high and Q_1 and the relay turn on when a resistance less than 10 MΩ appears across the touch contacts. If the touch contacts have a surface

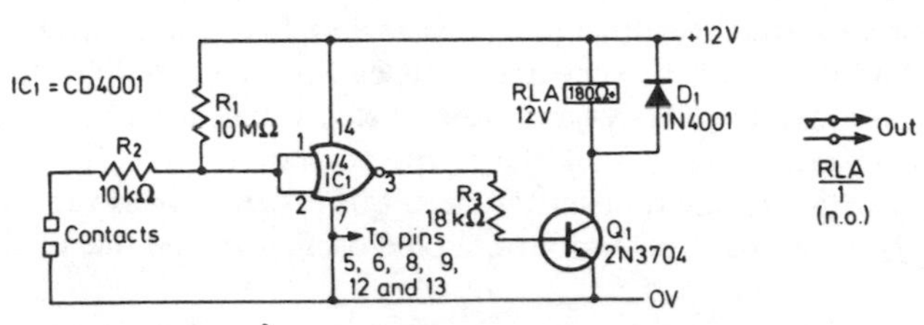

Figure 5.12 Resistive touch alarm, normal contacts

area of at least half a square centimetre each, a resistance of less than 10 MΩ will appear between them when they are simultaneously touched by an area of human skin, so the circuit acts effectively as a touch alarm. Any type of alarm device can be activated via the relay contacts, and the circuit consumes a typical standby current of only 1 μA.

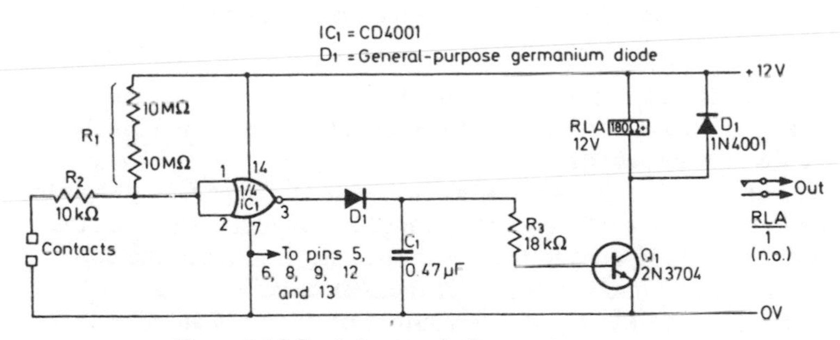

Figure 5.13 Resistive touch alarm, micro-contacts

Finally, *Figure 5.13* shows how the above circuit can be modified for use with micro-sized touch contacts. The circuit is similar to that described above, except that R_1 is increased to 20 MΩ by wiring two 10MΩ resistors in series, and that the design is also made sensitive to hum pick-up signals via D_1 and C_1. The sensitivity of this circuit is such that it can be used with pinhead-sized touch contacts.

An ultrasonic beam alarm

This unit can be used in the same type of application as the light-beam alarm circuits described in Chapter 4, but uses an invisible ultrasonic beam in place of a visible light beam. The circuit of the beam transmitter is shown in *Figure 5.14,* and the receiver/alarm is shown in *Figure 5.15.*

The circuits make use of an inexpensive matched pair of ultrasonic ceramic transducers of the type used in television remote-control units. These transducers are widely advertised in electronics magazines, and normally operate in the 30–50 kHz frequency range.

The operation of the transmitter circuit of *Figure 5.14* is very simple. Q_1 and Q_2 are wired as an emitter-coupled oscillator, and one of the transducers is used as the emitter coupling element, so the circuit oscillates at the basic frequency of the transducer and causes an ultrasonic signal to be generated.

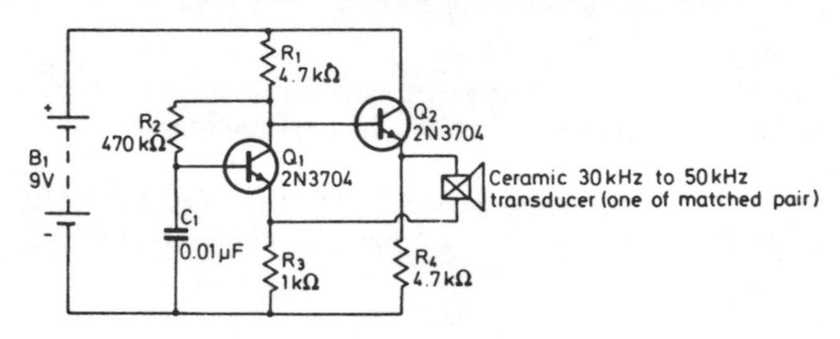

Figure 5.14 Ultrasonic beam alarm, transmitter

The operation of the receiver/alarm circuit of *Figure 5.15* is slightly more involved. Here, the remaining transducer is pointed towards the transmitter and responds to the transmitted signal in much the same way as a microphone. The output of the transducer is fed to tuned amplifier $Q_1-C_1-L_1-C_2$, and the output of the tuned amplifier is fed to an amplifying detector stage that is built around Q_2-D_1 and C_3. Normally, when the ultrasonic beam is unbroken, the output of this detector stage is high, so Q_3 is driven to saturation and Q_4 and the relay are cut off. When the beam is interrupted, the output of the detector stage falls to near zero volts, so Q_3 turns off and Q_4 and the relay are turned on via R_8. Any type of alarm device can be activated by the closing of the relay contacts. Thus the alarm is normally off, but turns on when the ultrasonic beam is interrupted.

The *Figure 5.15* circuit consumes a typical quiescent current of 5 mA, and the system has an operating range up to several yards. To set up the circuit, simply point the two transducers at one another over the required range, then carefully adjust R_4 so that 2 V d.c. are registered

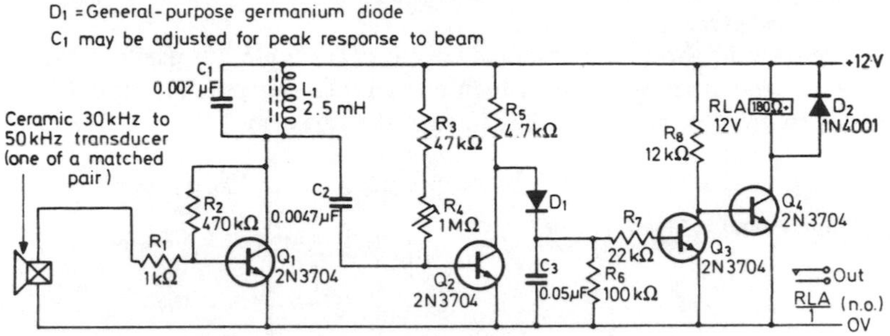

D_1 = General-purpose germanium diode
C_1 may be adjusted for peak response to beam

Figure 5.15 Ultrasonic beam alarm, receiver

across C_3 (on the 10 V range of a meter having a sensitivity of at least 20 kΩ/V) when the beam is uninterrupted. Then break the beam, and check that the voltage falls to near zero and the alarm turns on. If required, the value of C_1 can be adjusted to obtain optimum response at the ultrasonic operating frequency.

Sound and vibration alarm circuits

Sound-activated alarms can usefully be made to activate when an intruder enters a protected area and creates noise. Vibration-activated alarms can be made to activate when an unauthorised person opens the drawer of a cabinet or the door of a cupboard, etc., and thus creates a small amount of vibration in a protected object. Both types of circuit can use the same basic principle of operation, as illustrated in the block diagram of *Figure 5.16*.

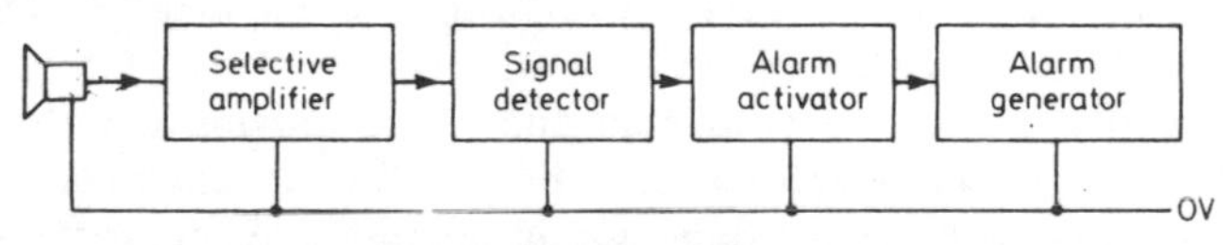

Figure 5.16 Block diagram of typical sound or vibration alarm circuit

Here, a microphone or similar transducer is used to pick up the basic noise or vibration of the environment, and the resulting signal is passed to a selective amplifier stage, which rejects unwanted signals and amplifies the band of signals that are of interest. The output of this amplifier is fed to a signal detector stage, which converts the a.c. input to a d.c. output; the d.c. output is fed to an alarm activator, which responds to input levels in excess of a predetermined amount. Finally, the output of the alarm activator is fed to the actual alarm signal generator.

Figure 5.17 shows the practical circuit of a simple but useful signal detector and alarm activator, which gives a relay output. The circuit needs an input of about 1 V a.c. to turn the relay on.

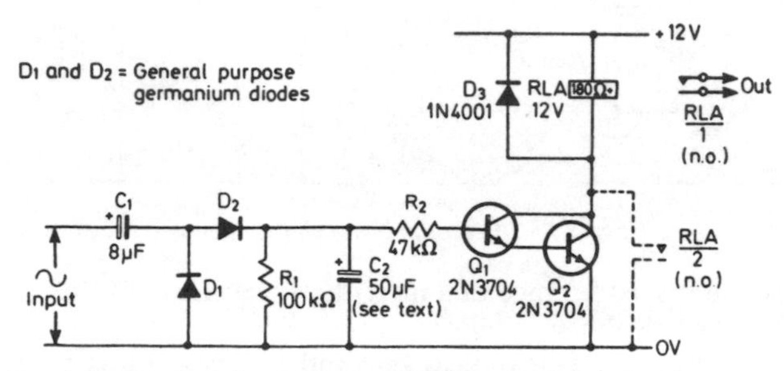

Figure 5.17 Simple relay-output signal-detecting alarm activator; needs 1 V a.c. input to activate relay

Here, the a.c. input signal is simply rectified and smoothed via D_1–D_2–R_1 and C_2, and the resulting d.c. is used to drive the relay on via Q_1 and Q_2. Normally, when zero input is applied, zero d.c. is developed across C_2, so Q_1–Q_2 and the relay are off. When an a.c. input of 1 V is fed to the circuit, roughly 2 V d.c. is developed across C_2, so Q_1–Q_2 and the relay are driven on. The circuit action is such that the relay turns on rapidly when a suitable input signal is connected, but turns off slowly when the signal is removed. The turn-off time is determined by the time constant of R_1 and C_2, and can be changed to suit specific requirements by altering the C_2 value.

The *Figure 5.17* circuit can be made self-latching by wiring a spare set of relay contacts across Q_2, as shown dotted in the diagram. The circuit can be used as a sound or vibration-activated alarm by feeding an a.c. input to it from a pick-up transducer via a suitable amplifier stage. In vibration-alarm applications, the amplifier should be designed to pass low-frequency signals only, and in sound-alarm applications it should be designed to pass the selected audio band only.

Finally, to conclude this chapter, *Figure 5.18* shows the circuit of an IC speech-frequency amplifier which can be used in conjunction with the *Figure 5.17* circuit to make a sensitive sound-activated alarm.

Here, sound is picked up by a 5 kΩ moving-coil microphone and is fed to pin 1 of the IC. The IC is a CA3035 ultra-high-gain wide-band amplifier array manufactured by RCA, and gives a voltage gain of about 120 dB between the input at pin 1 and the output at pin 7, so a vastly amplified version of the microphone signal appears at pin 7 and can be

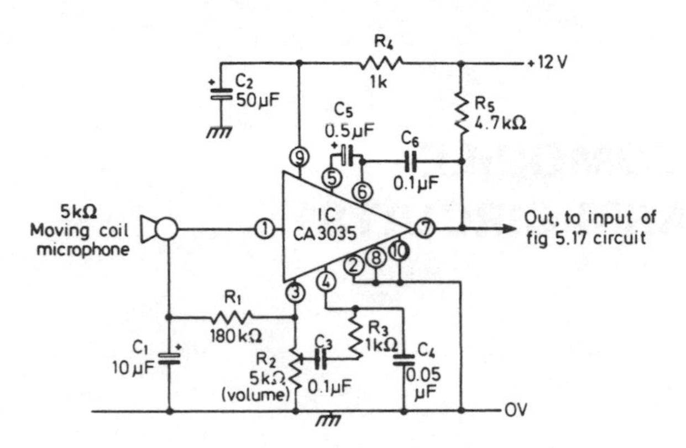

Figure 5.18 Speech-frequency amplifier; can be used in conjunction with Figure 5.17 to make a sound-activated alarm

fed to the input of the *Figure 5.17* circuit. R_1 and C_1 are bias components for the IC, and R_5 is a load resistor for one of the IC stages. R_4 $C_3-C_4-C_5$ and C_6 are coupling and frequency-compensation components for the IC, and R_5 is a load resistor for one of the IC stages, R_4 and C_2 provide a decoupled supply for two of the IC stages. The outline and pin numbers of the CA3035 IC are shown in the Appendix.

CHAPTER 6

AUTOMOBILE ALARM CIRCUITS

Electronic alarms have a number of practical applications in automobiles. They can be used to give anti-theft protection, or to indicate a probability of ice on the road or of overheating in the engine or gearbox, or to warn the driver that he is running low on fuel. Sixteen useful automobile alarm circuits are presented in this chapter. Included among these are five immobilisers which, although not true 'alarms', give useful anti-theft protection.

Types of anti-theft device

Vehicle anti-theft devices come in two basic types. The first of these is the 'immobiliser', which is intended simply to reduce a thief's chances of starting or driving away a target vehicle; it gives no protection against the car burglar, who merely wishes to steal objects that are left inside the vehicle. Immobilisers usually consist simply of a switch wired into some part of the electrical section of the vehicle's power unit, thus enabling the engine to be easily disabled.

The second type of anti-theft device is the true burglar alarm, which sounds an alarm (and perhaps also immobilises the vehicle's engine) if any unauthorised person tries to enter the vehicle. These alarms may be activated in one of three basic ways One of the most popular ways is via microswitches that operate when any of the car doors, hood (bonnet) or trunk (boot) is opened. Microswitch-activated alarms are fairly inexpensive, highly reliable, and can give excellent anti-theft protection.

Another way is by detecting the small drop that takes place in the vehicle's battery voltage when a door, hood or trunk courtesy-light turns on, or when the ignition is turned on. These so-called 'voltage sensing' alarms give the same degree of anti-theft protection as the microswitch types of alarm system, but are generally more expensive and less reliable.

A third way of activating an alarm is by detecting the vibration or swaying that takes place when a vehicle is entered or moved. This type of alarm has a number of disadvantages. If its sensitivity is adjusted so that it activates when anyone enters or rocks the vehicle, the system will tend to go off in gusty winds or when a person leans on the automobile. In this state the system has a very low reliability rating. Alternatively, if the system is adjusted so that it activates only when the vehicle is actually moved or subjected to substantial 'G' forces, it won't be sensitive enough to give effective anti-burglar protection. These systems can readily be made to give false alarms, so thieves can easily persuade their owners to disconnect them by repeatedly false-triggering the alarms.

Practical automobile anti-theft alarm systems can be switched on and off either from within the car or from outside. Systems that are switched from within the vehicle have a number of disadvantages. To enable the owner to leave the vehicle without activating the alarm, the system must incorporate a built-in 'exit' delay of about 30 seconds, and to enable the owner to enter the vehicle again it must have an additional built-in 'entry' delay of about 15 seconds.

Consequently the circuits tend to be fairly complex and expensive, and to have a relatively poor reliability rating. More important, the systems give very poor anti-burglar protection, since the thief is given a full 15 seconds of entry time in which to steal any worthwhile goodies before the alarm sounds off.

By contrast, externally-switched alarm systems can be very simple, reliable and inexpensive, and, since they can be made to sound off the instant that a car door starts to open, can be made to give excellent anti-burglar protection.

The comparative table of *Figure 6.1* shows the degree of protection offered by different types of anti-theft device against different types of thief. As can be seen, immobilisers give good protection against joy-riders and drive-away thieves, but give no protection against burglars or tow-away thieves, while externally-switched microswitch-activated and voltage-sensing alarms give good protection against all except tow-away thieves.

	Snatch burglar	*Cassette thief*	*Joy-rider*	*Drive-away thief*	*Tow-away thief*
Immobiliser	Nil	Nil	Good	Good	Nil
Internally-switched microswitch-activated alarm	Nil	Good	Good	Good	Nil
Internally-switched voltage-sensing alarm	Nil	Good	Good	Good	Nil
Internally-switched vibration alarm	Nil	Poor	Good	Poor	Fair
Externally-switched microswitch-activated alarm	Good	Good	Good	Good	Nil
Externally-switched voltage-sensing alarm	Good	Good	Good	Good	Nil
Externally-switched vibration alarm	Poor	Poor	Good	Poor	Fair

Figure 6.1 Comparative table showing degree of protection given by different types of anti-theft device against different types of thief

Having cleared up these points, let's now go on and look at some practical anti-theft circuits.

Immobiliser circuits

Immobilisers simply reduce a thief's chances of starting or driving away a target vehicle. Simple immobilisers consist of a concealed switch wired into some part of the electrical section of the vehicle's power unit. *Figures 6.2* to *6.5* show a number of circuits of this type.

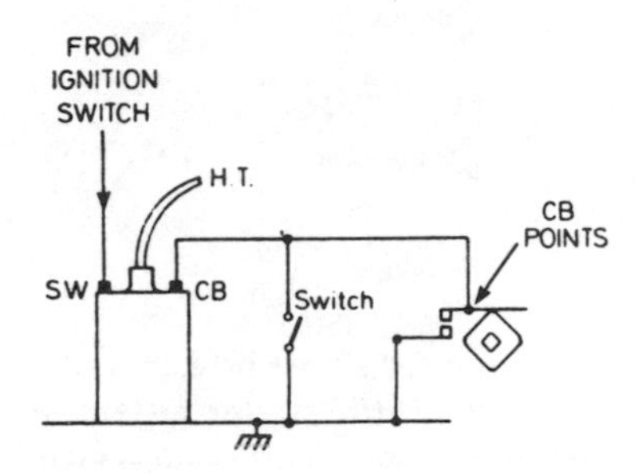

Figure 6.2 Contact-breaker immobiliser, operates when switch is closed

Figures 6.2 and *6.3* show how immobilisers can be wired into the vehicle's ignition system. In *Figure 6.2*, the switch is wired across the vehicle's contact-breaker (CB) points. When the switch is open the ignition operates normally, but when the switch is closed the CB points

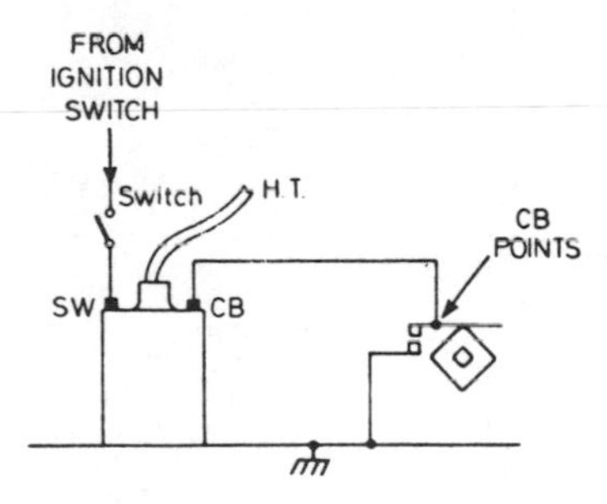

Figure 6.3 Ignition immobiliser, operates when switch is open

are shorted out and the engine is unable to operate. This circuit gives excellent protection, particularly if the wiring is carefully concealed at the CB end.

In the *Figure 6.3* circuit the immobiliser switch is wired in series with the vehicle's ignition switch, so that the engine operates only when the switch is closed. The protection of this circuit is not as good as that of *Figure 6.2,* since a skilled thief can by-pass the immobiliser and ignition switches by simply hooking a wire from the battery to the SW terminal of the coil.

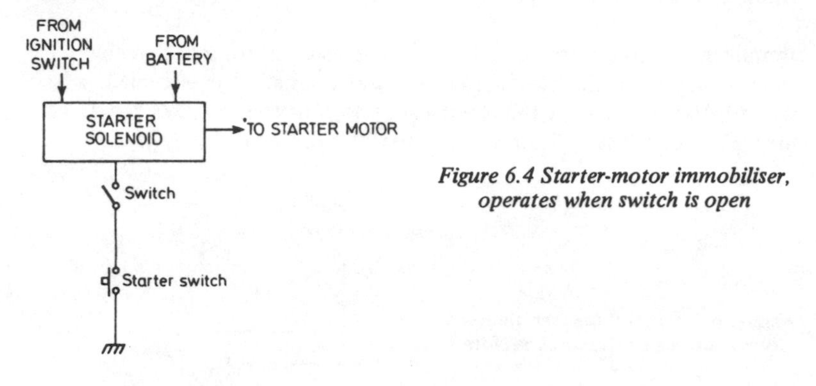

Figure 6.4 Starter-motor immobiliser, operates when switch is open

Figure 6.4 shows how an immobiliser switch can be wired into the vehicle's electric starter system, so that the starter only operates if this switch is closed. This system gives better protection than *Figure 6.3,* but is not as good as *Figure 6.2* because the starter solenoid can be operated manually on many vehicles, and also because the starter and immobiliser switches can be by-passed by a single length of wire.

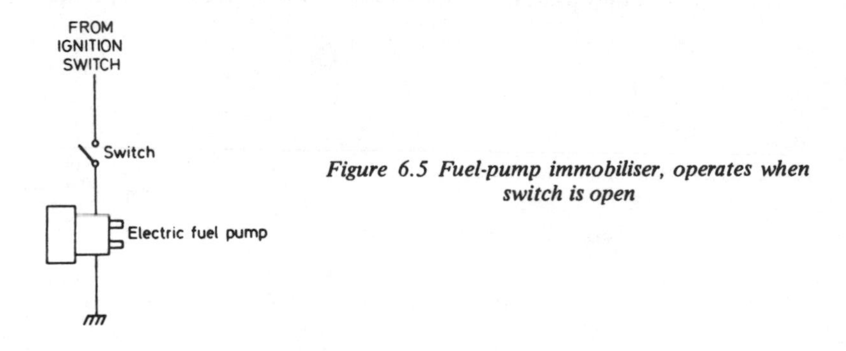

Figure 6.5 Fuel-pump immobiliser, operates when switch is open

Finally, *Figure 6.5* shows how an immobiliser switch can be wired in series with the electric fuel pump on suitable vehicles, so that the pump operates only when this switch is closed. A feature of this system is that it permits a thief to start the engine and drive for a short distance on the fuel remaining in the carburettor before the lack of fuel-pump operation immobilises the vehicle.

A weakness of the *Figure 6.2* to *6.5* circuits is that they must all be turned on and off manually, so they only give protection if the owner remembers to turn them on. By contrast, *Figure 6.6* shows an immobiliser that turns on automatically when an attempt is made to start the engine, but that can be turned off by briefly operating a hidden push-button

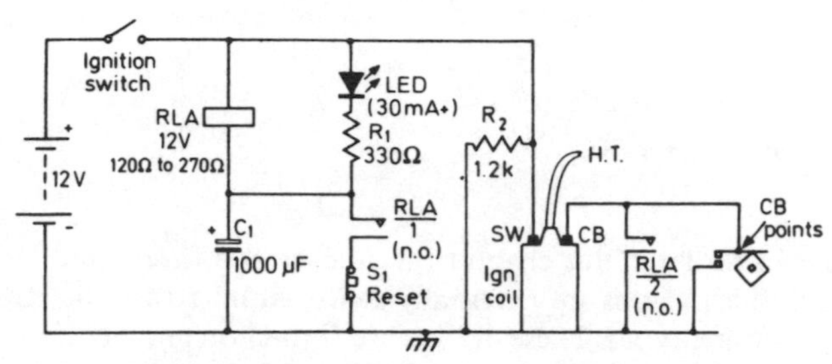

Figure 6.6 Self-activating immobiliser circuit for negative-ground vehicles; for positive-ground vehicles reverse the polarities of C_1 *and LED*

switch. A small 'reminder' light turns on when the engine is disabled by the immobiliser. This circuit thus gives a high degree of protection, since it does not depend on the memory of its owner. The circuit operates as follows.

The coil of relay *RLA* is wired in series with 1000 μF capacitor C_1, and the combination is wired across the vehicle's ignition switch. C_1 is shunted by the series combination of n.o. relay contacts $RLA/1$ and n.c. push-button switch S_1; n.o. relay contacts $RLA/2$ are wired across the vehicle's CB points; an LED (light-emitting diode) is wired in series with current-limiting resistor R_1, and the combination is wired across the relay coil.

Normally C_1 is fully discharged. Consequently, when the ignition switch is first closed a surge of current flows through the relay coil via C_1, and the relay turns on. As the relay goes on, contacts $RLA/1$ close and lock the relay on via S_1, and contacts $RLA/2$ close and short out the vehicle's CB points, thus immobilising the engine. Under this condition current flows in the LED via R_1, and the LED illuminates. The relay stays on until S_1 is briefly opened, at which point the relay unlatches and C_1 charges up rapidly via the relay coil, and the relay and the LED turn off. As the relay turns off, the short is removed from the vehicle's CB points, and the engine is able to operate in the normal way.

The relay used in the *Figure 6.6* circuit can be any 12 V type with a coil resistance in the range 120 Ω to 270 Ω, and with two or more sets of n.o. contacts. The LED can be any type with a mean current rating greater than 30 mA. The circuit as shown is for use on vehicles with negative-ground electrical systems. On positive-ground vehicles, reverse the polarities of C_1 and the LED.

Anti-theft alarm circuits

It was shown earlier in this chapter that the most efficient and useful vehicle anti-theft alarms are externally-switched microswitch-activated or voltage-sensing types. These alarms are turned on and off via a concealed toggle-switch or a prominent key-switch fitted to the outside of the vehicle. *Figures 6.7* to *6.10* show practical examples of alarm systems of these types. All these circuits also act as immobilisers, operating the vehicle's horn and lights and immobilising the engine under the 'alarm' condition.

In the *Figure 6.7* to *6.9* circuits, microswitches that are built into the vehicle are used to trip a self-latching relay when any of the car doors, hood or trunk is opened; this relay immobilises the engine and operates the horn and headlights either directly or via additional circuitry. Two suitable front-door microswitches are built into most vehicles as standard fittings, and are used to operate the courtesy or dome lights. Additional switches can easily be fitted to the rear doors. The hood and trunk can be protected by 'auxiliary' microswitches.

The operation of the *Figure 6.7a* circuit is very simple. Normally, with the key-switch open, no voltage is fed to the relay network, so the alarm is off. Suppose, however, that the key-switch is closed. If any of the door switches close, current flows in the relays via D_1; if any of the auxiliary switches close, current flows via D_2. In either case, both relays turn on. As *RLA* goes on, contacts *RLA*/1 close and lock both relays on, and contacts *RLA*/2 close and short out the vehicle's CB points, thus immobilising the vehicle.

Simultaneously contacts *RLB*/1 close and switch on the car horn, giving an audible indication of the intrusion, and contacts *RLB*/2 close and switch on the headlights, giving a visual identification of the violated vehicle. The horn and lights remain on until the key-switch is opened, or until the vehicle's battery runs flat.

The *Figure 6.7a* circuit is for use on negative-ground vehicles. The circuit can be modified for use on positive-ground vehicles by simply reversing the polarities of D_1 and D_2, as shown in *Figure 6.7b*.

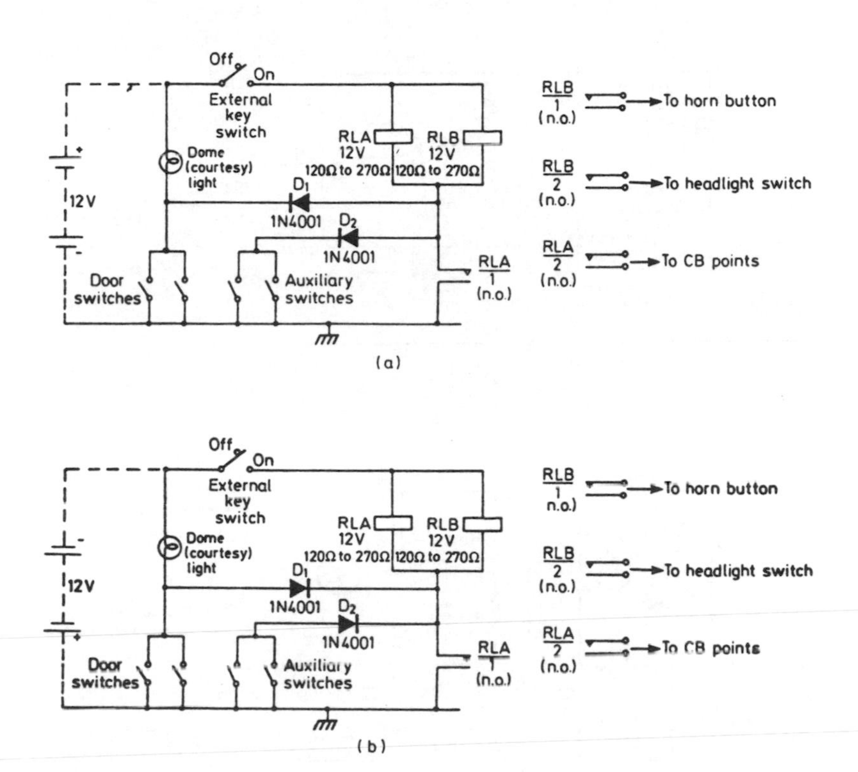

Figure 6.7 Simple microswitch-activated anti-theft alarm/immobiliser, operates horn and lights until switched off or until battery runs flat; (a) negative-ground, (b) positive-ground

A weakness of the simple *Figure 6.7* circuit is that, since car horns and their associated components are not designed to withstand continuous long-period operation, these components may be damaged if the alarm sounds for too long. *Figure 6.8a* shows how the *Figure 6.7a* circuit can be modified so that the horn and lights turn off automatically after four minutes or so, thus minimising the possibility of horn damage.

Here, *RLA* turns on and self-latches in the same way as in the *Figure 6.7a* circuit, and as contacts *RLA*/1 close the full battery voltage is applied across the Q_1-Q_2-RLB network. At the moment that power is applied, C_1 is fully discharged and acting like a short-circuit, so the

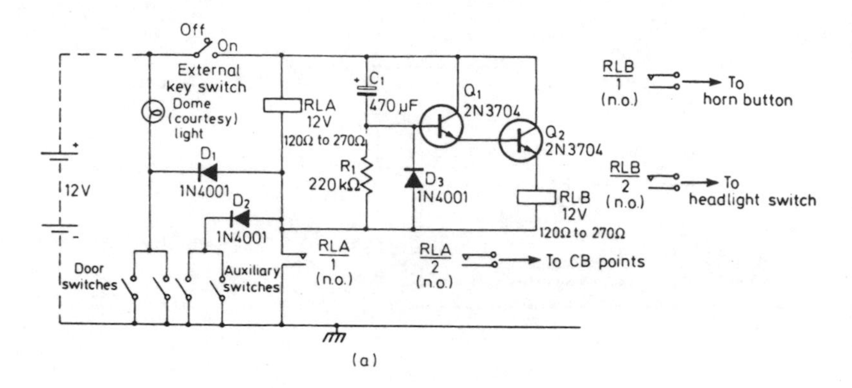

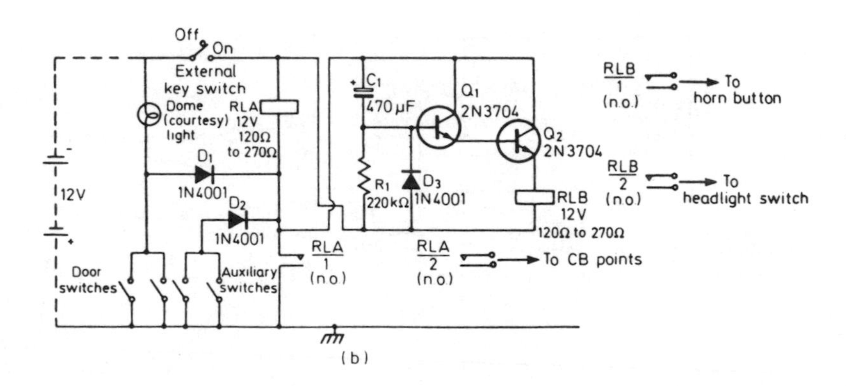

Figure 6.8 Improved microswitch-activated anti-theft alarm/immobiliser, turns horn and lights off automatically after four minutes; (a) negative-ground, (b) positive-ground

base and collector of Q_1 are effectively shorted together. *RLB* is thus immediately turned on via the Q_1-Q_2 Darlington emitter-follower, and the horn and lights operate.

As soon as the power is applied to the circuit, C_1 starts to charge up via R_1, and the voltage across the coil of *RLB* starts to decay exponentially towards zero. After a delay of about four minutes this voltage falls so low that *RLB* and the horn and lights turn off. *RLA* remains on, however, until the system is turned off via the key-switch, so the vehicle remains immobilised via its CB points.

The *Figure 6.8a* circuit is for use on negative-ground vehicles. The circuit can be modified for use on positive-ground vehicles by reversing the polarities of D_1 and D_2, and reversing the supply connections to the *RLB*-driving network, as shown in *Figure 6.8b*.

A minor practical snag with the *Figure 6.8* circuit is that, since it gives a 'monotone' form of horn operation, its owner is unlikely to be

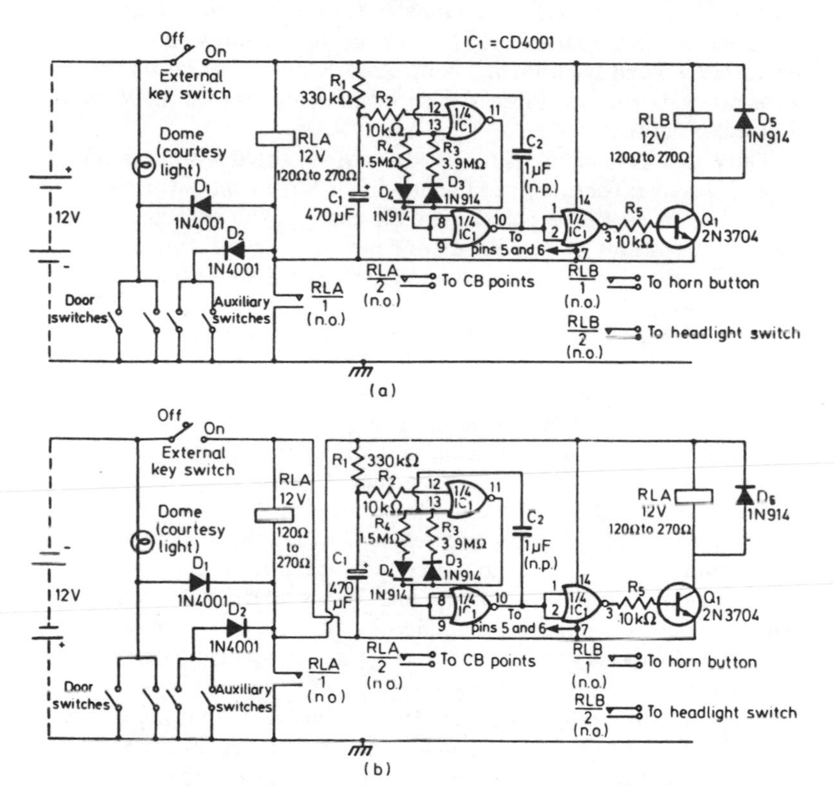

Figure 6.9 Modified microswitch-activated anti-theft alarm/immobiliser, gives distinctive 'pulsed' operation of horn and lights and turns them off automatically after four minutes; (a) negative-ground, (b) positive-ground

able to recognise the sound of his own vehicle, and will tend to check his own vehicle whenever he hears any horn sound off. This snag is overcome in the circuit of *Figure 6.9a*, which pulses the horn and lights on for 4 seconds and off for 1.5 seconds repeatedly for about four minutes under the alarm condition, thus producing a very distinctive warning signal.

The *Figure 6.9a* circuit is similar to that of *Figure 6.8a,* except that *RLB* is driven by a pulse generator formed from Q_1 and a type CD4001 COS/MOS digital IC. The IC is wired as a buffered-output gated astable multivibrator, with unequal 'on' and 'off' times, and with its gating controlled by time-delay network R_1-C_1. The 'on' time of the relay is controlled by R_3-D_3 and approximates 4 seconds; the 'off' time is controlled by R_4-D_4 and approximates 1.5 seconds. Note that C_2 is a non-polarised (n.p.) capacitor. The pulse generator turns on and activates *RLB* and the horn and lights as soon as *RLA* turns on, but turns off again automatically after about four minutes via the R_1-C_1 time-delay network.

The *Figure 6.9a* circuit is for use on vehicles fitted with negative-ground systems. The circuit can be modified for use on positive-ground vehicles by reversing the polarities of D_1 and D_2, and reversing the supply connections to the *RLB*-driving network, as in *Figure 6.9b.*

Finally, *Figure 6.10a* shows the practical circuit of a voltage-sensing type of alarm, which can be used in place of the simple *RLA*-driving network described in the earlier circuits. Circuit operation relies on the fact that a small but sharp drop occurs in battery voltage whenever a vehicle courtesy light, etc., is turned on. This sudden drop in voltage is detected and made to operate *RLA*. The system has the advantage that the alarm's pick-up can be attached directly to the vehicle's battery, rather than to a number of microswitches.

The operation of the *Figure 6.10a* circuit is fairly simple. Here, potential divider $R_1-R_2-R_3$ is wired across the vehicle's supply lines. The output of this divider is fed directly to the inverting (pin 2) terminal of an open-loop type 741 op-amp, but is taken to the non-inverting (pin 3) terminal via a simple ($R_4-C_1-R_5$) time-delay or memory network. A small 'offset' voltage can be applied between the input terminals of the op-amp via R_6.

Suppose then that the offset control is adjusted so that the voltage of pin 2 is fractionally higher than that of pin 3 under 'steady voltage' conditions, and that under this condition the output of the op-amp is driven to negative saturation. If now a small but abrupt fall occurs in the supply voltage, this fall is transferred immediately to pin 2 of the op-amp, but does not immediately reach pin 3 because of the time-delay or memory action of C_1. Consequently, pin 2 briefly goes negative relative to pin 3, and as it does so the output of the op-amp is driven briefly to positive saturation, thus giving a positive output pulse. This pulse is used to charge C_2 via D_1, and C_2 drives Q_1-Q_2 and the relay on via R_8. As the relay goes on, contacts $RLA/1$ close and cause the relay to self-latch, and contacts $RLA/2$ close and immobilise the vehicle via its CB points.

Note that the above circuit responds only to sudden drops in potential, and is not influenced by absolute values of battery voltage. Thus leaving the car lights, etc., on or off has no influence on the operation of the alarm system.

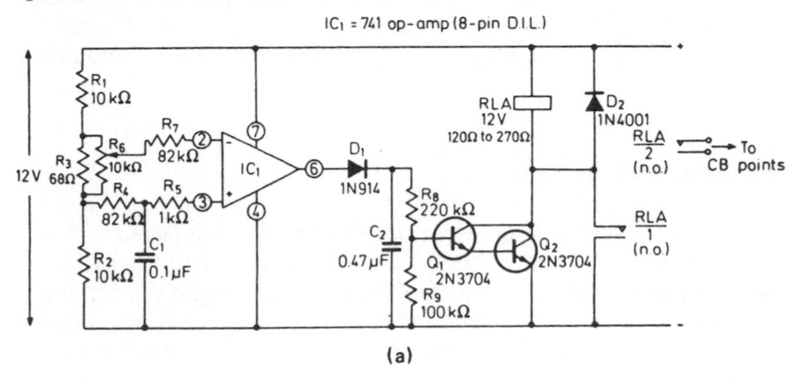

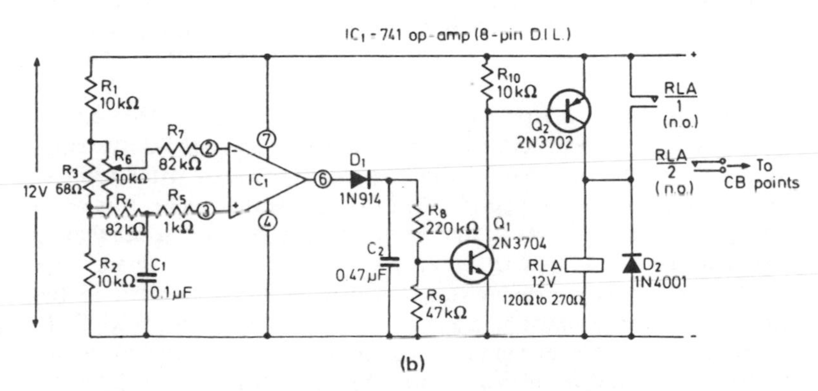

Figure 6.10 Voltage-sensing alarm circuit, can be used to replace the RLA-driving network in alarm circuits described earlier; (a) negative-ground, (b) positive-ground

The *Figure 6.10a* circuit is intended for use on negative-ground vehicles, and can be used directly in place of the *RLA* network in any of the *Figure 6.7a, 6.8a,* or *6.9a* circuits. The circuit can be modified for use on positive-ground vehicles by using the connections shown in *Figure 6.10b,* and can then be used directly in place of the *RLA* network in any of the *Figure 6.7b, 6.8b,* or *6.9b* circuits.

When installing the *Figure 6.10* circuit in a vehicle, R_6 must be adjusted so that the alarm turns on reliably when a courtesy light goes on, without being excessively sensitive to small shifts in battery voltage. To find the correct R_6 setting, proceed as follows.

First remove the courtesy lamp and replace it with one having half the original current rating. Now adjust R_6 just past the point where the alarm fails to operate when the lamp goes on, and then turn R_6 back a fraction, so that the alarm only just operates via the courtesy light. Finally refit the original courtesy lamp. Reliable operation should then be obtained.

Installing anti-theft alarms

The anti-theft alarms described in this chapter are all designed to be turned on and off via an externally mounted switch. This 'on/off' control can take the form of a carefully concealed toggle switch, or a prominently mounted key-operated switch. In either case, the switch should be mounted so that neither it nor its wiring is vulnerable to damage by weather, road dirt or potential car thieves. If a key-operated switch is used it should be mounted in a prominent position, close to the driver's door, so that it acts as a visual deterrent to potential car thieves.

Once the alarm's master 'on/off' switch has been fitted, the next installation job is to fit suitable microswitches to activate the system. As already mentioned, two suitable switches are already fitted to most vehicles, and are used to operate the dome or courtesy light. It is worth fitting additional switches to the rear doors, and essential to fit them to the trunk and hood if full anti-theft protection is to be obtained. Note that if your vehicle is fitted with a voltage-sensing type of alarm system, these microswitches must be made to switch a lamp or similar kind of current load. The higher the load current used, the more reliable will be the operation of the alarm circuit. The microswitches can all be wired in parallel, and a single load used.

Finally, when the installation is complete, give your system a complete functional check. When conducting this test, try not to annoy your neighbours.

Ice-hazard alarms

Ice-hazard alarms activate when the vehicle's ignition is turned on and the air temperature a little way above the road surface is at or below

0°C. The alarms thus indicate a hazard of meeting ice under actual driving conditions.

Two useful ice-hazard alarm circuits are shown in this section. In each case the circuits act as precision under-temperature alarms, and use a thermistor as a temperature sensor. The thermistor is mounted in the air-flow at the front of the vehicle, a little way above the road surface, and gives a good indication of the actual road temperature. Both circuits are based on designs already presented in Chapter 3.

The first circuit, shown in *Figure 6.11*, gives a relay-output warning of the ice-hazard, and is designed around a type 741 operational

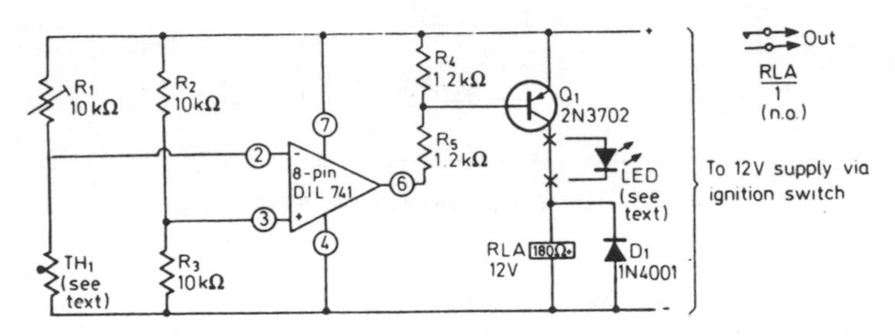

Figure 6.11 Relay-output ice-hazard alarm

amplifier. The design is based on that of *Figure 3.12*, and a full description of the circuit operation is given in Chapter 3. The following additional points should, however, be noted about the *Figure 6.11* circuit.

The relay used in the circuit can be any 12 V type with one or more sets of n.o. contacts and having a coil resistance of 180 Ω or greater. Any type of external alarm-indicator can be activated via the relay contacts.

If required, an LED can be made to activate when the alarm turns on, thus giving a visual indication of the alarm condition. When relays with coil resistances of 300 Ω or greater are used, the LED can be any type with a rating of 40 mA or greater, and can be wired in series with the relay coil as indicated in the circuit diagram. Alternatively, if a coil resistance less than 300 Ω is used, a more sensitive LED can be wired in series with a suitable current-limiting resistor and the combination can be wired in parallel with the relay coil.

The second circuit, shown in *Figure 6.12*, gives an 800 Hz pulsed-output loudspeaker warning of the ice-hazard, and is designed around a

CD4001 COS/MOS IC. The design is based on that of *Figure 3.15a,* and a full description of the circuit operation is given in Chapter 3.

The speaker used in the *Figure 6.12* circuit can have any impedance in the range 3 Ω to 100 Ω, and the value of R_X should be chosen to give a combined series impedance of 100 Ω. If required, an LED with

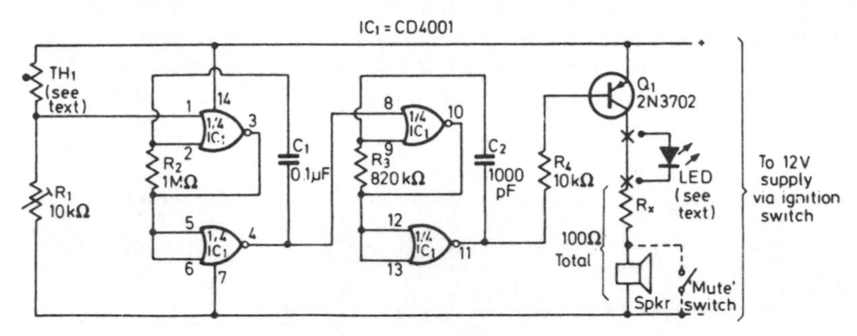

Figure 6.12 800 Hz pulsed-output ice-hazard alarm

a current rating of at least 120 mA peak or 60 mA mean can be wired in series with R_X (as in the diagram) to give a visual indication of the alarm condition. A useful addition to the *Figure 6.12* circuit is a muting switch, to reduce the speaker volume once the alarm call has been noted. The volume can be reduced to zero by wiring the switch directly across the speaker as shown, or can be reduced to a preset level by wiring a limiting resistor (value found by trial and error) in series with the switch.

The power connections of both ice-hazard alarm circuits should be taken to the vehicle's battery via the ignition switch, so that the circuits are automatically energised whenever the vehicle is in use. On vehicles that use an ignition dropper resistor, the connections can be taken to the battery via an ignition-switch-activated relay contact.

The thermistors used in the two circuits can be any negative-temperature-coefficient types that present a resistance in the range 1 kΩ to 10 kΩ at 0°C. Each thermistor must be mounted in a small 'head' that is fixed to the lower front of the vehicle, and connected to the main alarm-unit via twin flex. To make the thermistor head, solder the thermistor to a small tag-board and solder its leads to the twin flex. Coat the whole assembly with waterproof varnish, so that moisture will not affect its apparent resistance, then mount it in a small plastic or metal box and fix it to the lower front of the vehicle. Before fixing the head in place, however, calibrate the alarm system as follows.

Immerse the head in a small container filled with a water and ice mixture. Use a thermometer to measure the temperature of the mixture, and add ice until a steady reading of 0° C is obtained. Now adjust R_1 so that the alarm just turns on; raise the temperature slightly, and check that the alarm turns off again. If satisfactory, the head and the alarm system can now be fixed permanently to the vehicle.

Overheat-warning alarms

Each of the ice-hazard alarms of *Figures 6.11* and *6.12* can be modified so that it activates when its thermistor temperature goes above (rather than below) a preset value. In such cases the thermistor can be used as a probe that can be bonded to any fixed part of the vehicle. The circuits can thus be used to warn the driver of overheating in the engine, gearbox, differential, brake drums, etc. Two practical overheat-warning alarm circuits are presented in this section.

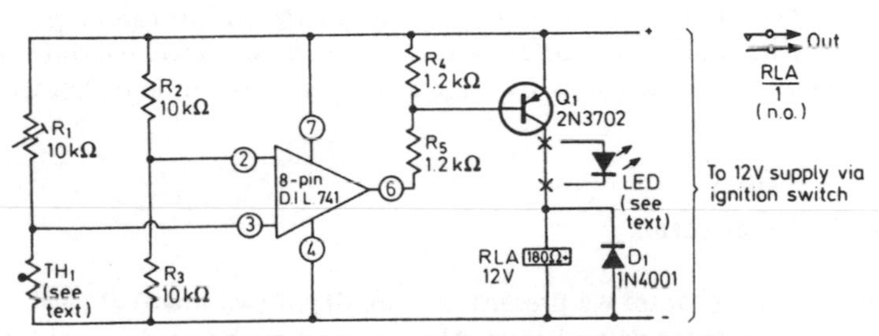

Figure 6.13 Relay-output overheat-warning alarm

The first circuit, shown in *Figure 6.13*, gives a relay-output warning of the overheat condition, and is identical to the *Figure 6.11* ice-hazard alarm, except that the pin 2 and pin 3 input connections of the op-amp are transposed, so that the alarm activates when the temperature goes above (rather than below) a preset value.

The second circuit, shown in *Figure 6.14*, gives an 800 Hz pulsed-output loudspeaker warning of the overheat condition, and is identical to the *Figure 6.12* ice-hazard alarm, except that the positions of R_1 and the thermistor are transposed, so that the alarm activates when the temperature goes above a preset value.

The notes applying to the use of LEDs, muting switches, relay types and thermistor types, etc., in the section on ice-hazard alarms also apply to the two overheat circuits of *Figures 6.13* and *6.14*.

To set up either circuit, raise its thermistor to the desired overheat alarm temperature, and adjust R_1 so that the alarm just turns on. Then

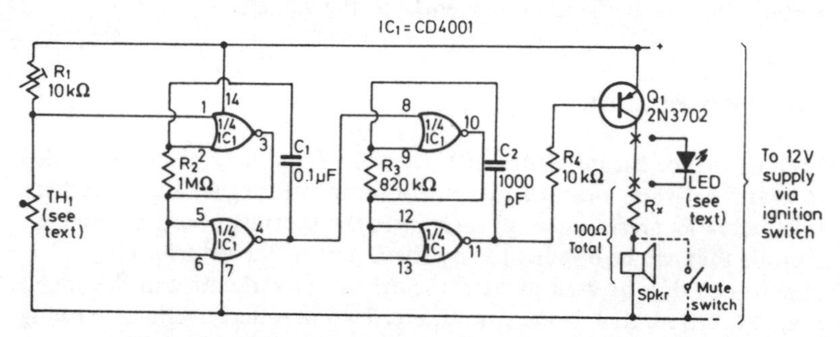

Figure 6.14 800 Hz pulsed-output overheat-warning alarm

reduce the temperature slightly, and check that the alarm turns off again. When the circuit is correctly set up, its thermistor can be permanently bonded to the surface whose temperature is to be monitored (i.e. engine, gearbox, brake drums, etc.), using epoxy resin or 'plastic metal'.

Low-fuel-level alarms

Finally in this chapter we present the circuits of two low-fuel-level alarms, which give relay outputs. These alarms can be used on vehicles fitted with a 12 V electrical system and with a fuel gauge of the type that is actuated via a fuel-tank-mounted potentiometer, in which the voltage developed across the potentiometer is proportional to the fuel level, i.e. the voltage decreases as the fuel level falls. The alarms need a minimum input (from the potentiometer) of 1.5 V for satisfactory operation.

To find out if the alarms are suitable for use in your own vehicle, simple measure the voltage across the tank-mounted potentiometer, or between the 'low' terminal of the fuel gauge and ground: check that a steady voltage reading is obtained, roughly proportional to the fuel level, and greater than 1.5 V under the required low-fuel-level alarm condition.

The circuit of the negative-ground version of the alarm is shown in *Figure 6.15*. Here, Q_1 and Q_2 are wired as a simple differential voltage amplifier, with its output feeding to relay-driving transistor Q_3, and Q_2

and Q_3 are wired together as a regenerative switch, with backlash controlled via R_5 and R_6. One input of the differential amplifier is derived from the fuel gauge via D_1–R_1 and C_1, which form a simple smoothing network and ensure that the Q_1 base voltage corresponds to the mean

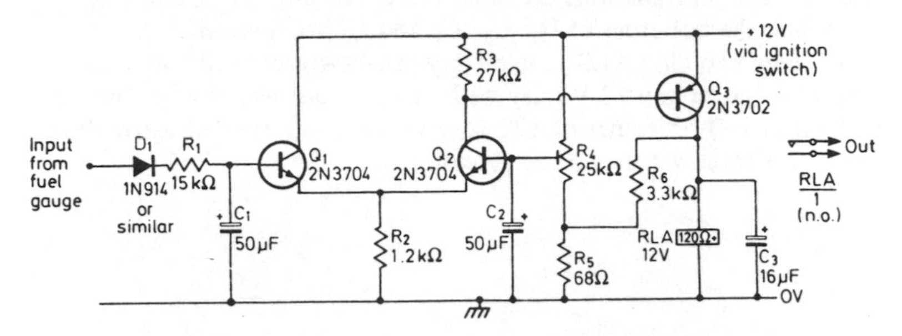

Figure 6.15 Relay-output low-fuel-level alarm, negative-ground

(rather than instantaneous) input voltage. The other input is derived from the 12 V supply line via R_4, which sets a reference voltage on the base of Q_2. C_2 and C_3 ensure that neither the amplifier nor the relay is influenced by supply-line transients or rapid changes in the vehicle's battery voltage.

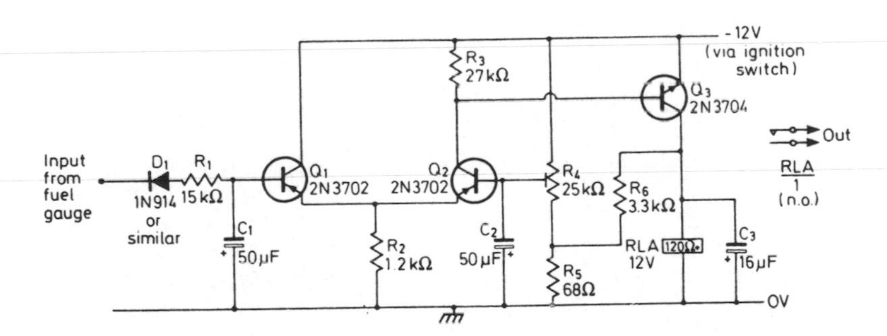

Figure 6.16 Relay-output low-fuel-level alarm, positive-ground

The action of the circuit is such that the relay is normally off, but turns on when the voltage on the base of Q_1 (from the fuel gauge) falls below that of Q_2 (the reference voltage). Once the relay has turned on, the value of the reference voltage is automatically increased via the regenerative action of Q_2 and Q_3. Thus the alarm turns on as soon as the voltage from the fuel-gauge (and thus the fuel level) falls below a preset value, but once it has turned on the relay does not turn off again until the fuel level rises appreciably above the initial trigger level.

Finally, the positive-ground version of the low-fuel-level alarm is shown in *Figure 6.16.* This circuit is identical to that of *Figure 6.15,* except that npn transistors are used in place of pnp types, and vice versa, and the polarities of D_1, C_1, C_2 and C_3 are reversed.

In these two circuits D_1 can be any general-purpose silicon diode, and *RLA* can be any 12 V relay with one or more sets of n.o. contacts and with a coil resistance of 120 Ω or greater. Any type of alarm device can be activated via the n.o. relay contacts.

CHAPTER 7

INSTRUMENTATION ALARM CIRCUITS

Instrumentation alarms can be used to activate a lamp or other visual indicator when a monitored voltage or resistance, etc., goes beyond preset limits. In this chapter we present a dozen simple but useful instrumentation alarm circuits that can be used to monitor a.c. or d.c. voltages or currents, or resistance. Most of the circuits are designed around an 8-pin d.i.l type 741 operational amplifier.

All the circuits are designed to give an LED (light-emitting diode) output, thus giving a visual indication of the alarm condition. This LED can be any type having a mean current rating up to 40 mA, and must be wired in series with a current-limiting resistor, shown as R_y in the circuit diagrams. R_y must have its value chosen to match the current rating of the LED to the supply voltages that are used with each circuit. The formula for finding the value of this resistor is

$$R_y = \frac{V_{supply} - 2}{I_{LED}}$$

Thus, if the circuit has a 12 V supply and the LED is a 40 mA (= 0.040 A) type, R_y must be given a value of roughly 250 Ω. Note that if a circuit uses two sets of supplies (such as +9 V and −9 V), V_{supply} must be taken as the difference between the two supply voltages (= 18 V in this example). In the circuits, R_y is shown as having a nominal value of 330 Ω, and this value should be close enough to 'correct' for most practical purposes.

D.C. voltage alarm circuits

Figure 7.1 shows the practical circuit of a precision d.c. over-voltage alarm, which works with inputs in excess of 5 V only. Here, the op-amp is used in the open-loop mode as a d.c. voltage comparator, with a zener-derived 5 V reference signal applied to the non-inverting pin of

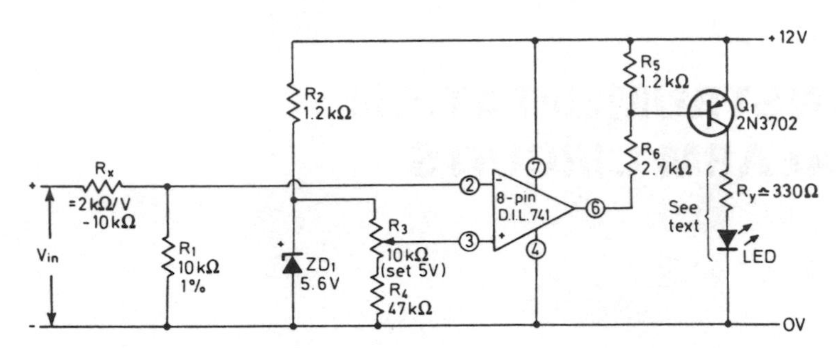

Figure 7.1 Precision d.c. over-voltage alarm, covering 5 V upwards

the op-amp via R_3, and the test voltage applied between the inverting pin and ground. The circuit action is such that the op-amp output is positively saturated, and Q_1 and the LED are off, when the inverting pin voltage is fractionally less than the 5 V reference potential, and Q_1 and the LED go on when the inverting pin voltage is fractionally greater than the 5 V reference potential.

R_X is wired in series between the input test voltage and the 10 kΩ impedance of the inverting pin of the op-amp, and enables the circuit to be ranged so that it triggers at any required voltage in excess of the 5 V reference value. The R_X value for any required trigger voltage is determined on the basis of 2 kΩ/V – 10 kΩ. Thus, for 50 V triggering, $R_X = (50 \times 2\ \text{k}\Omega) - 10\ \text{k}\Omega = 90\ \text{k}\Omega$. For 5 V triggering, R_X must have a value of zero ohms.

The *Figure 7.1* circuit is very sensitive and exhibits negligible backlash. Triggering accuracies of 0.5 per cent can easily be achieved. For maximum accuracy, either the power supply or the zener reference voltage of the circuit should be fully stabilised.

The *Figure 7.1* circuit can be made to function as a precision under-voltage alarm, which turns on when the input voltage falls below a preset level, by transposing the inverting and non-inverting pin connections of the op-amp, as shown in *Figure 7.2.* This circuit also shows how the zener reference supply can be stabilised for high-precision operation.

Note in both of these circuits that, once 5 V has been accurately set via R_3, the final triggering accuracy of each design is determined solely by the accuracies of R_X and R_1. In high-precision applications, therefore, these resistors should be precision wire-wound types.

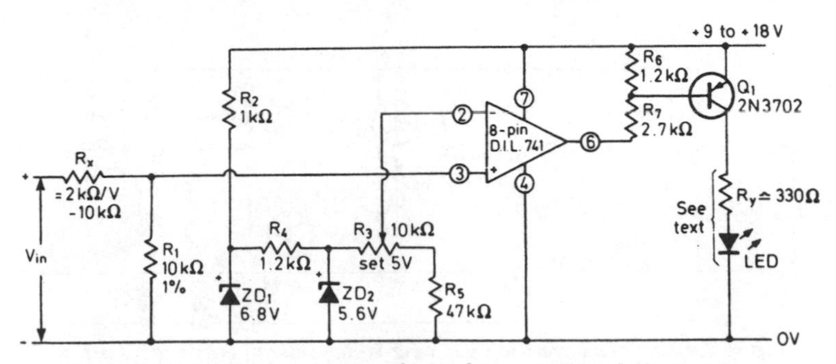

Figure 7.2 Precision d.c. under-voltage alarm, covering 5 V upwards

Figure 7.3 shows how the *Figure 7.1* circuit can be modified for use as an over-voltage alarm covering the range 10 mV to 5 V. In this case the input voltage is connected directly to the inverting terminal of the op-amp, and a variable reference potential is applied to the non-inverting terminal. This reference potential is adjusted to give the same value as

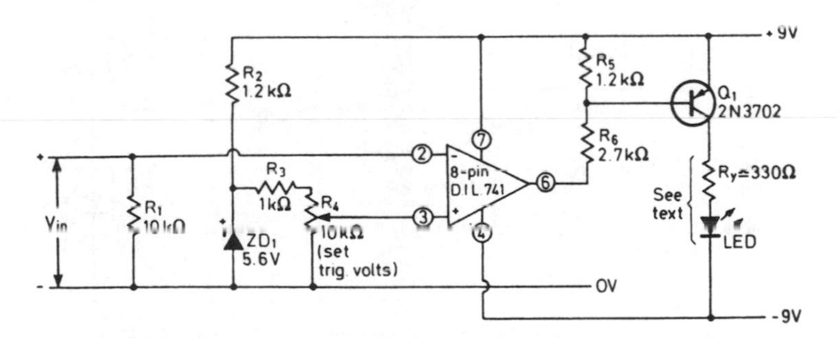

Figure 7.3 Dual-supply precision d.c. over-voltage alarm, covering 10 mV to 5 V

that of the required trigger voltage. The circuit action can be reversed, so that the design acts as an under-voltage alarm, by transposing the input pin connections of the op-amp. Note that the *Figure 7.3* circuit uses two sets of supply lines, to ensure proper biasing of the op-amp.

Figure 7.4 shows how the *Figure 7.3* circuit can be adapted for operation from a single set of supply lines. Here Q_2 and Q_3 are wired as an astable multivibrator or square-wave generator, and the output of this generator is used to provide a negative supply rail for the op-amp via voltage-doubling and smoothing network D_1–D_2 and C_3–C_4. The

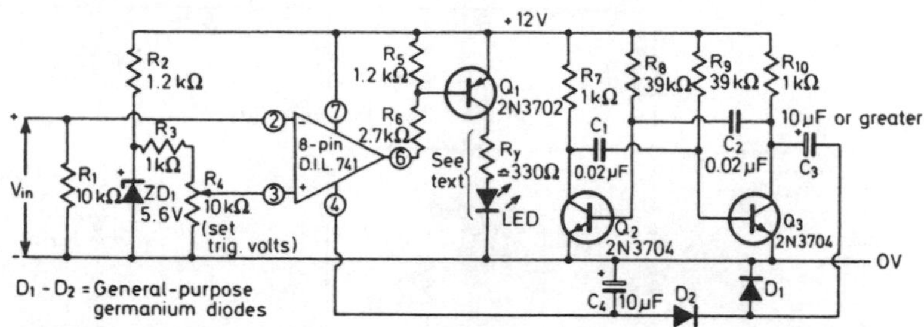

Figure 7.4 Single-rail precision d.c. over-voltage alarm, covering 10 mV to 5 V

doubler gives a negative output of about 9 V when unloaded, but gives only 3–5 volts when connected to pin 4 of the op-amp.

Finally, *Figure 7.5* shows the circuit of a d.c. over-voltage alarm that covers the range 10 mV to 5 V and uses a single floating supply. Here, the op-amp is again used as a d.c. voltage comparator, but its positive

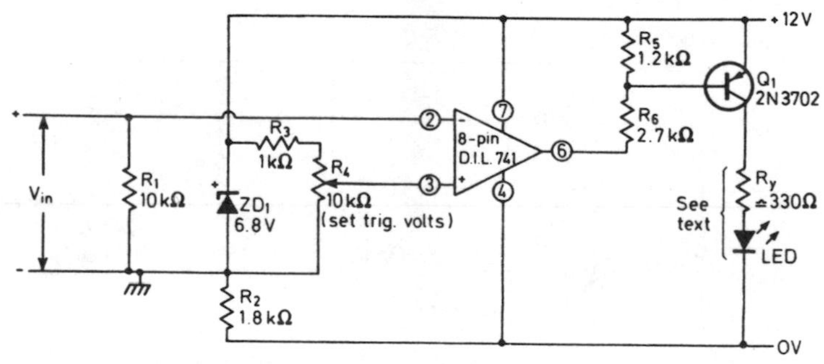

Figure 7.5 Single-supply d.c. over-voltage alarm, covering 10 mV to 5 V

supply rail is set at 6.8 V via the floating supply and zener diode ZD_1, and its negative rail is set at –5.2 V via the ZD_1 and R_2 combination. The monitored input signal is fed to the inverting terminal of the op-amp, and a zener-derived reference potential is fed to the non-inverting terminal via R_3 and R_4. This reference potential can be varied between roughly 10 mV and 5 V, and this is therefore the voltage range covered by the over-voltage alarm.

A.C. voltage alarm circuits

The five voltage-activated alarms shown in *Figures 7.1* to *7.5* are designed for d.c. activation only. All these circuits can be modified for a.c. activation by interposing suitable rectifier/smoothing networks or

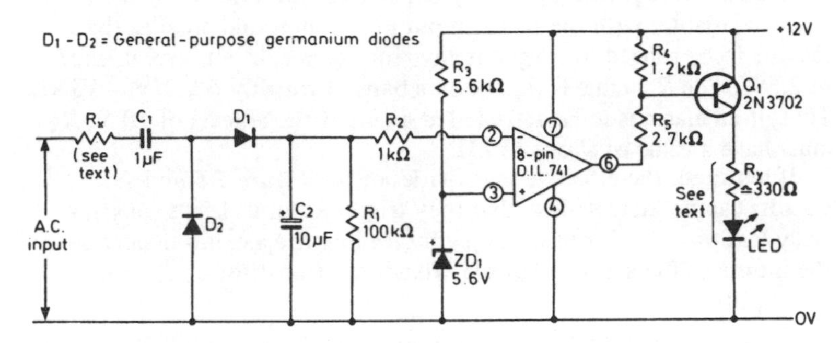

Figure 7.6 Precision a.c. over-voltage alarm, covering 2.5 V upwards

a.c./d.c. converters between their input terminals and the actual a.c. input signals, so that the a.c. signals are converted to d.c. before being applied to the alarm circuits.

Figure 7.6 shows the practical circuit of a precision a.c. over-voltage alarm that is designed to work with sine-wave signals in excess of

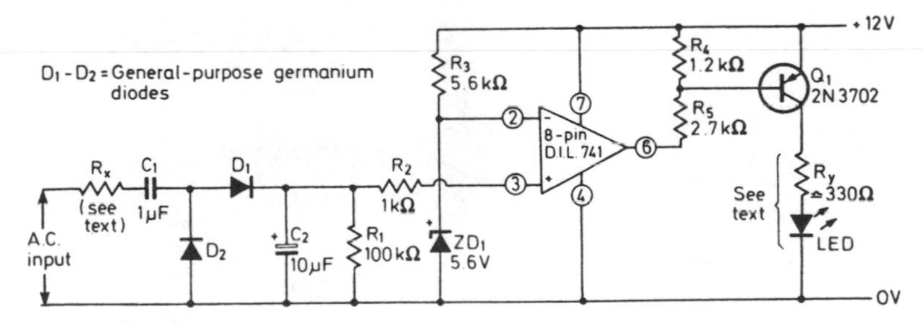

Figure 7.7 Precision a.c. under-voltage alarm, covering 2.5 V upwards

2.5 V r.m.s. Here, the a.c. signal is converted to d.c. via voltage doubling and smoothing network R_X–C_1–D_1–D_2–C_2–R_1, and the resulting d.c. voltage is applied to the inverting input of the op-amp via R_2. A zener-derived 5.6 V reference potential is applied to the non-inverting terminal. The circuit action is such that the alarm turns on when the d.c. voltage on the inverting terminal exceeds 5.6 V.

The action of the above circuit can be reversed, so that it works as an under-voltage alarm, by transposing the input terminal connections of the op-amp, as shown in *Figure 7.7.*

The circuits of *Figures 7.6* and *7.7* both exhibit a basic input impedance, with R_X reduced to zero ohms, of about 15 kΩ, and under this condition a sine wave of about 2.5 V r.m.s. is needed to activate the alarm. Consequently, when R_X is given a finite value it acts as a potential divider with the 15 kΩ input impedance, and enables the circuits to be ranged to trigger at any required a.c. input level in excess of 2.5 V. The R_X value is chosen on a basis of roughly 6 kΩ/V – 15 kΩ. Thus, if an alarm is to be activated at an input signal level of 10 V, R_X must have a value of about 45 kΩ.

If required, the effective sensitivities of the *Figure 7.6* and *7.7* circuits can be increased, so that they trigger at input levels substantially less than 2.5 V r.m.s., simply by feeding the a.c. input signals to the inputs of the alarm circuits via fixed-gain transistor or op-amp preamplifiers.

Current alarm circuits

Each of the five d.c. voltage alarm circuits of *Figures 7.1* to *7.5* can be used as a d.c. current alarm by simply feeding the monitored current to the input of the voltage alarm via a current-to-voltage converter. A suitable converter circuit is shown in *Figure 7.8.*

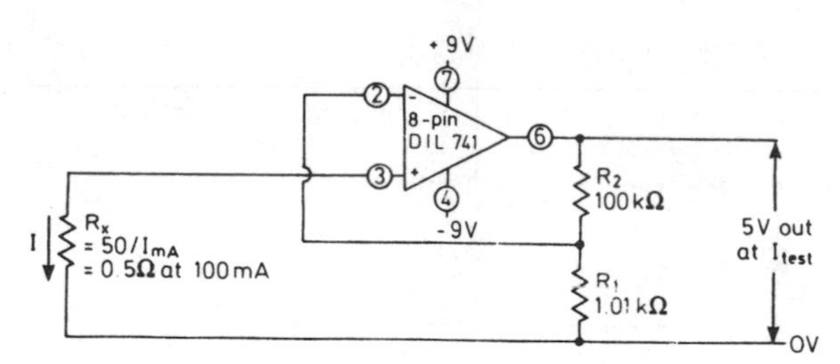

Figure 7.8 D.C. current-to-voltage converter

Here, the op-amp is wired as a non-inverting x 100 voltage amplifier, with gain determined by the ratios of R_1 and R_2. The test current is passed through input resistor R_X, which has its value chosen so that

50 mV is developed across it at the required trigger current, thus giving 5 V output from the op-amp under this condition. The R_X value is selected on the basis of

$$R_X = \frac{50}{I_{mA}}$$

where I_{mA} is the desired trigger current in milliamps. Thus R_X needs a value of 0.5 Ω at trigger levels of 100 mA, or 0.05 Ω at trigger levels of 1 A.

A similar type of converter circuit, using a.c. coupling, can be used to enable the *Figure 7.6* and 7.7 circuits to act as a.c. current alarms.

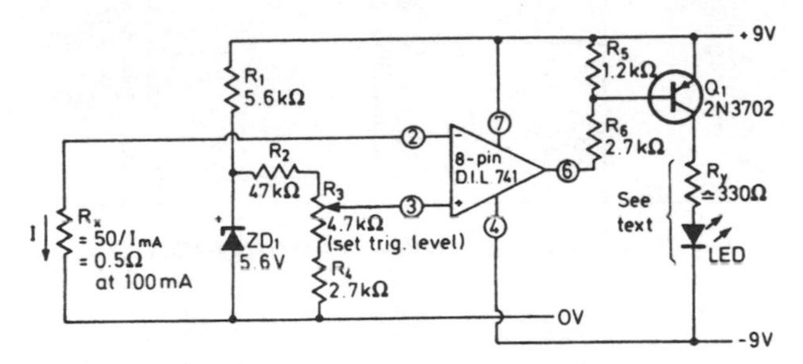

Figure 7.9 D.C. over-current alarm

If required, an op-amp circuit can be built specifically for use as a d.c. current alarm. *Figure 7.9* shows an over-current version of such a circuit. Here, R_X is again used to develop a potential of 50 mV at the desired current level, and this voltage is applied to the inverting pin of the op-amp. A zener-derived reference potential of approximately 50 mV is applied to the non-inverting terminal of the op-amp. This reference voltage can be adjusted over a limited range via R_3, thus providing a limited control of the circuit's sensitivity.

Thus the *Figure 7.9* alarm circuit turns on when the current-derived input voltage exceeds the 50 mV potential of the reference voltage. The action of the circuit can be reversed, so that it acts as an under-current alarm, simply by transposing the connections to the two input terminals of the op-amp. In either case, the value of monitor resistor R_X is chosen on the basis of

$$R_X = \frac{50}{I_{mA}}$$

Resistance alarms

Figure 7.10 shows the practical circuit of a precision under-resistance alarm, which turns on when the value of a monitored resistance falls below a preset value. Here, the op-amp is again used as a voltage

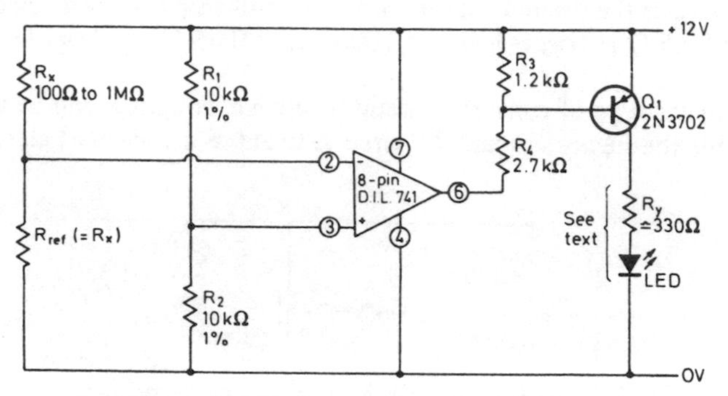

Figure 7.10 Precision under-resistance alarm

comparator, with its output feeding to the LED visual indicator via Q_1, but in this case the voltage to the non-inverting pin of the op-amp is set at half-supply volts via potential divider R_1–R_2, and the voltage on the inverting pin is determined by the ratios of R_X and R_{ref}. In effect, these four resistors are wired as a Wheatstone bridge, and the circuit action is such that the alarm turns on when the value of R_X falls below that of reference resistor R_{ref}, i.e. when the bridge goes out of balance in such a way that the voltage on the inverting terminal rises above that of the non-inverting terminal of the op-amp.

R_X and R_{ref} must have equal values, but can be given any values in the range 100 Ω to 1 MΩ. The minimum resistance value is dictated by the current-driving capability of the circuit's power supply, and the maximum value is restricted by the shunting effect that the input of the op-amp has on the effective value of R_{ref}.

The accuracy of the above circuit is independent of variations in power-supply voltage, and the alarm is capable of responding to changes of less than 0.1 per cent in the value of R_X. The actual accuracy is determined by R_1–R_2 and R_{ref}, and in worst-case terms is equal to the sum of the tolerances of these three resistors, i.e. it equals ±3 per cent if 1 per cent resistors are used.

The action of the above circuit can be reversed, so that it acts as a precision under-resistance alarm, by transposing the input pin connections of the op-amp, as shown in *Figure 7.11*. This circuit also shows

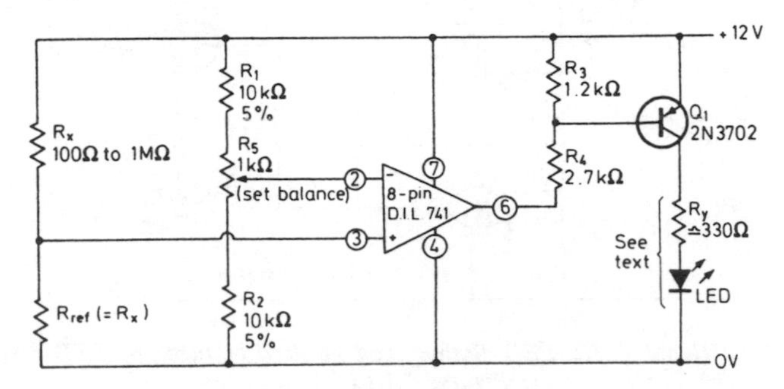

Figure 7.11 Precision over-resistance alarm

how the accuracies of both designs can be improved by adding R_5 'set balance' control to the R_1–R_2 divider chain. This control enables the bridge to be precisely balanced, so that the alarm turns on when the value of R_x varies from the marked value of R_{ref} by only 0.1 per cent or so. In this case the true accuracy of the circuit is equal to the tolerance of R_{ref} plus 0.1 per cent.

An LED flasher circuit

All of the circuits that we have looked at so far in this chapter are intended to be built into electronic instruments, and have a simple LED output that is intended to be fitted to the front panel of the instrument and to turn on and illuminate under the alarm condition. Each of these circuits can be modified, if desired, so that the LED flashes on and off rapidly under the alarm condition, thus giving a more attention-catching indication that a fault has occurred. *Figure 7.12* shows the circuit of the LED flasher, together with details of how it can be added to the alarm circuits.

Here, two of the gates of a CD4001 COS/MOS digital IC are wired as a gated 6 Hz astable multivibrator, and the output of this astable is used to drive the LED via R_3 and Q_1. Normally, when the pin 1 terminal of the IC is high, the astable is disabled and the LED is off. When pin 1 goes low, on the other hand, the astable is gated on and pulses the LED on and off at a rate of roughly 6 Hz. The rate can be altered by changing the R_2 value.

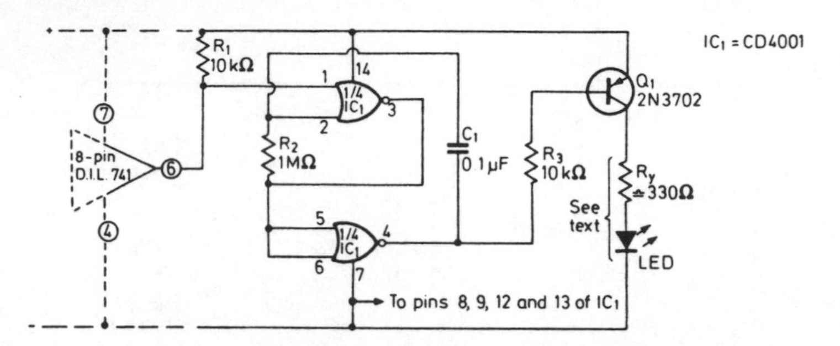

Figure 7.12 Gated 6 Hz LED flasher, can be used in place of LED driver in Figures 7.1 to 7.11

The LED flasher circuit uses the existing power supplies of the alarm circuits, and is used in place of the existing LED driver networks. Pin 1 of the COS/MOS IC is simply connected directly to output pin 6 of the op-amp of the alarm circuit. *Note:* the supply voltage between pins 7 and 14 of the IC must not exceed 18 V.

CHAPTER 8

INFRA-RED LIGHT-BEAM ALARMS

In the Chapter 4 survey of light-sensitive circuits we briefly mentioned infra-red (IR) light-beam systems which can be used as the basis of high-security intrusion detectors/alarms. In this final chapter we expand on this theme by showing ways of making practical IR invisible-light-beam alarms. We start off, however, by looking at some basic principles.

Intrusion alarm basics

A simple invisible-light-beam intrusion detector or alarm system can be made by connecting an IR light transmitter and receiver as shown in *Figure 8.1*. Here, the transmitter feeds a coded signal (often a simple square wave) into an IR LED which has its output focused into a fairly narrow beam (via a moulded-in lense in the LED casing) that is aimed at a matching IR photo-transistor in the remotely placed receiver. System action is such that the receiver output is 'off' when the light-beam reaches the receiver, but turns 'on' and activates an external relay or alarm if the beam is interrupted by a person or animal, etc. This basic type of system can be designed to give an effective detection range up

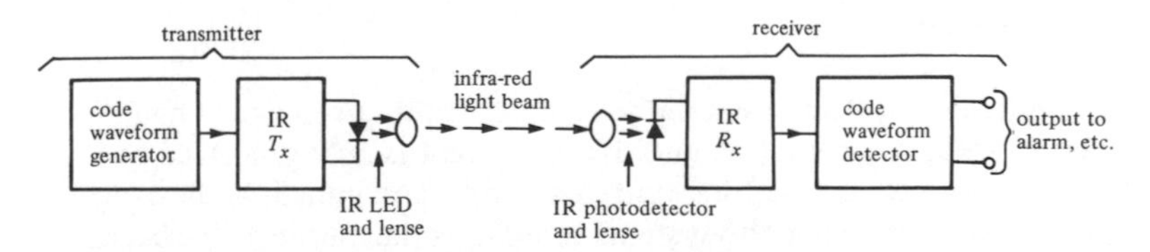

Figure 8.1 Simple light-beam intrusion alarm/detector system

to 30 metres when used with additional optical focusing lenses, or up to 8 metres without extra lenses.

The above system works on the pin-point 'line-of-sight' principle and can be activated by any 'bigger-than-a-pin' object that enters the line-of-sight between the transmitter and receiver lenses. Thus, a weakness of this simple system is that it can be false-triggered by a fly or moth (etc.) entering the beam or landing on one of the lenses. The improved 'dual-light-beam' system of *Figure 8.2* does not suffer from this defect.

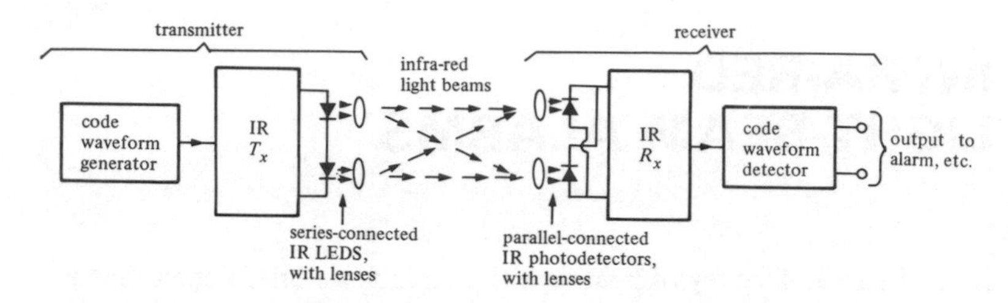

Figure 8.2 Dual-light-beam intrusion alarm/detector system

The *Figure 8.2* system is basically similar to that already described, but transmits the IR beam via two series-connected LEDs that are spaced about 75 mm apart, and receives the beam via two parallel-connected photodetectors that are also spaced about 75 mm apart. Thus, each photodetector can detect the beam from either LED, so the receiver activates only if *both* beams are broken simultaneously, and this will normally only occur if a large (greater than 75 mm) object is placed within the composite 'beam'. This system is thus virtually immune to false triggering via moths, etc.

Note that, as well as giving excellent false-alarm immunity, the dual-light-beam system also gives (at any given LED drive-current value) double the effective detection range of the simple single-beam system (i.e. up to 16 metres without additional lenses), since it has twice as much effective infra-red transmitter output power and twice the receiver sensitivity.

System waveforms

Infra-red beam systems are usually used in conditions in which high levels of ambient or 'background' IR radiation (usually generated by heat sources such as radiators, tungsten lamps, and human bodies, etc.) already exist. To enable the systems to differentiate against this background radiation and give good effective detection ranges, the transmitter beams are invariably frequency modulated, and the receivers are fitted with matching frequency detectors. In practice, the transmitted beams invariably use either continuous-tone or tone-burst frequency modulation, as shown in *Figure 8.3*.

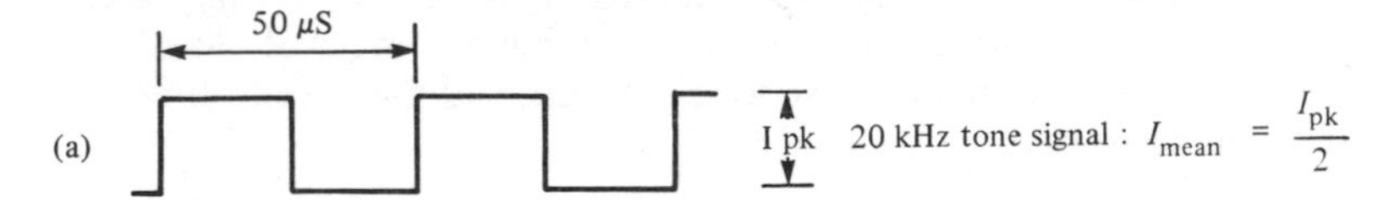

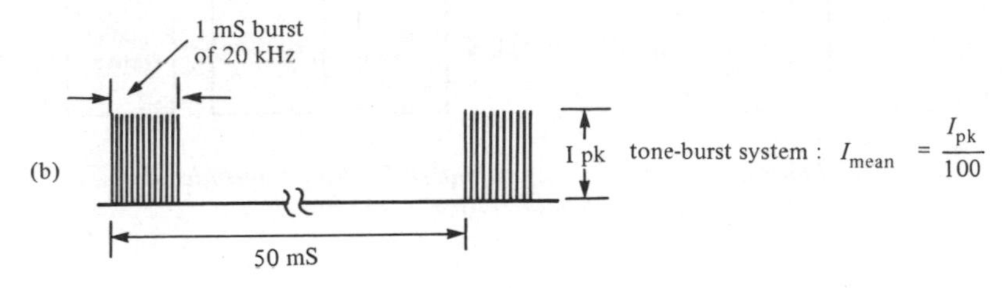

Figure 8.3 Alternative types of IR light-beam code waveforms, with typical parameter values

Infra-red LEDs and photodetectors are very fast acting devices, and the effective range of an IR beam system is thus determined by the peak current fed into the transmitting LED, rather than by its mean LED current. Thus, if the waveforms of *Figure 8.3* are used in transmitters giving peak LED currents of 100 mA, both systems will give the same effective operating range, but the *Figure 8.3a* continuous-tone transmitter will consume a mean LED current of 50 mA, while the tone-burst system of *Figure 8.3b* will consume a mean current of only 1 mA (but will require more complex circuit design).

The operating parameters of the tone-burst waveform system require some consideration, since the system actually works on the 'sampling' principle. For example, it is a fact that at normal walking speed a human takes about 200 mS to pass any given point, so a practical IR light-beam burglar alarm system does not need to be turned on continuously, but only needs to be turned on for brief 'sample' periods at repetition periods far less than 200 mS (at, say, 50 mS); the 'sample' period should be short relative to repetition time, but long relative to the period of the 'tone' frequency. Thus, a good compromise is to use a 20 kHz tone with a burst or sample period of 1 mS and a repetition time of 50 mS, as shown in *Figure 8.3b*.

System design

The first step in designing any electronics system is that of drawing up suitable block diagrams; *Figure 8.4* shows a suitable block diagram of a continuous-tone IR intrusion alarm/detector system, and that of *Figure 8.5* shows that of a tone-burst system. Note that a number of blocks (such as the IR output stage, the tone pre-amp, and the output driver) are common to both systems.

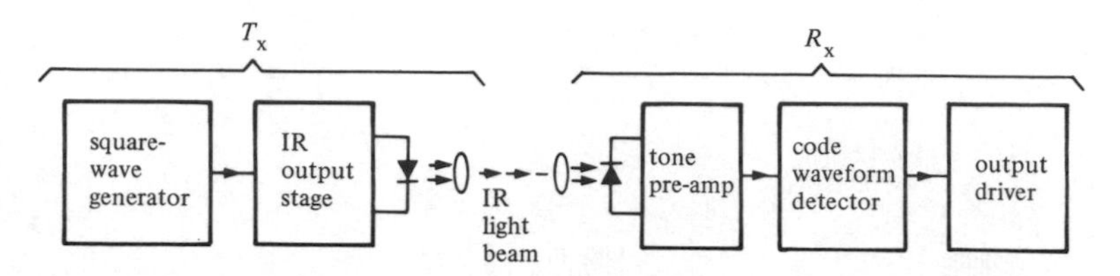

Figure 8.4 Block diagram of continuous-tone IR light-beam intruder alarm/detector system

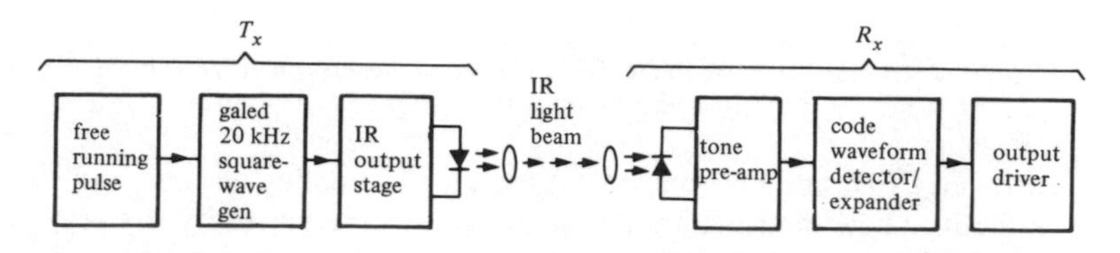

Figure 8.5 Block diagram of tone-burst IR light-beam intruder alarm/detector system

The continuous-tone system (*Figure 8.4*) is very simple, with the transmitter comprising nothing more than a square-wave generator driving an IR output stage, and the receiver comprising a matching tone pre-amplifier and code waveform detector, followed by an output driver stage that activates devices such as relays and alarms, etc.

The tone-burst system is far more complex, with the transmitter comprising a free-running pulse generator (generating 1 mS pulses at 50 mS intervals) that drives a gated 20 kHz square-wave generator, which in turn drives the IR output stage, which finally generates the tone-burst IR light beam. In the receiver, the beam signals are picked up and passed through a matching pre-amplifier, and are then passed on to a code waveform detector/expander block, which ensures that the alarm does not activate during the 'blank' parts of the IR waveform. The output of the expander stage is fed to the output driver.

Figure 8.6 shows an alternative version of the tone-burst system.

This is similar to the above, except that a simple code waveform detector is used in the receiver section, and that a 'blanking' gate is interposed between the detector and the output driver and is directly driven by the transmitter's pulse generator, to ensure that the alarm is not activated during the blank parts of the IR waveform.

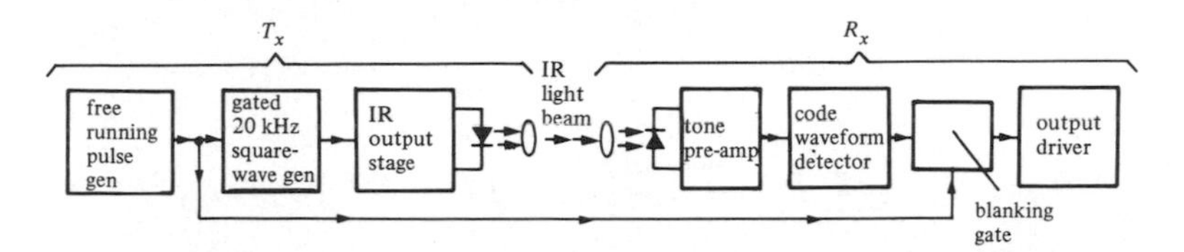

Figure 8.6 Block diagram of alternative IR light-beam system

Transmitter circuits

Figure 8.7 shows the practical circuit of a simple continuous-tone dual-light-beam IR transmitter. Here, a standard '555 timer' IC is wired as an astable multivibrator that generates a non-symmetrical 20 kHz square-wave output that drives the two IR LEDs at peak output currents of about 400 mA via R_4 and Q_1 and the low source impedance of storage capacitor C_1. The timing action is such that the 'on' period of the LEDs is controlled by C_2 and R_2, and the 'off' period by C_2 and $(R_1 + R_2)$, i.e., so that the LEDs are 'on' for only about one-eighth of each cycle; the circuit thus consumes a *mean* current of about 50 mA.

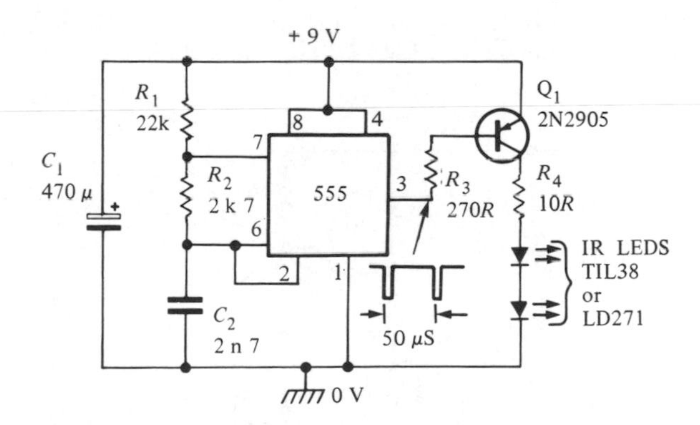

Figure 8.7 Simple continuous-tone IR light-beam transmitter

The above circuit can use either TIL38 or LD271 (or similar) high-power IR LEDs. These devices can handle mean currents up to only

150 mA, but can handle brief repetitive peak currents several times greater than this value. *Figure 8.8* shows the outline and connections of these devices, which have a moulded-in lense that focuses the output into a radiating 'beam' of about 60 degrees width; at the edges of this beam the IR signal strength is half of that at the centre of the beam.

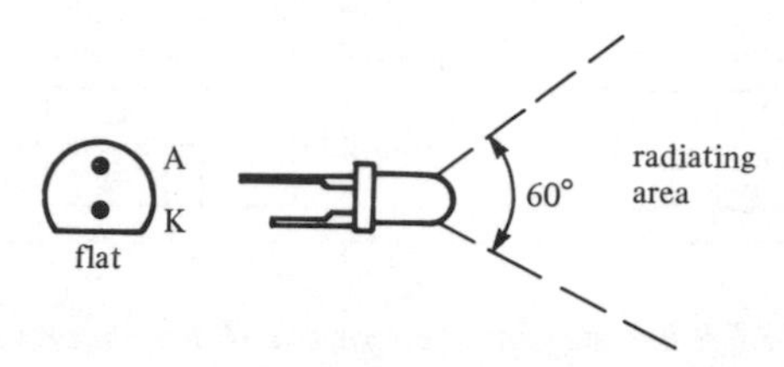

Figure 8.8 Outline and connections of the LD271 and TIL38 IR LED

Minor weaknesses of the IR output stage (Q_1 and R_3-R_4) of the *Figure 8.7* circuit are that it has a very low input impedance (about 300 Ω), that it gives an inverting action (the LEDs are on when the input is low), and that the LED output current varies with the circuit supply voltage. *Figure 8.9* shows an alternative 'universal' IR transmitter output stage that suffers from none of these defects.

In *Figure 8.9*, the base drive current of output transistor Q_2 is derived from the collector of Q_1, which has an input impedance of about 5k0 (determined mainly by the R_1 value). Thus, when the input is low Q_1 is off, so Q_2 and the two IR LEDs are also off, but when the input is high Q_1 is driven to saturation via R_3, thus driving LED_1 (a standard red LED) and Q_2 and the two IR LEDs on.

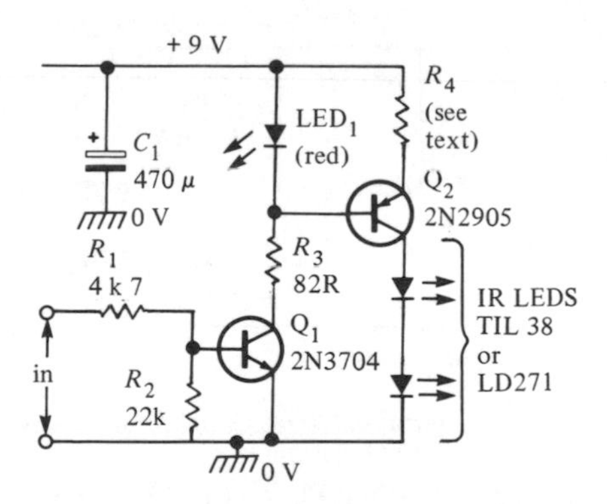

Figure 8.9 'Universal' IR T_x output stage

Note that under this latter condition about 1.8 V are developed across LED_1, and that about 0.6 V less than this (= 1.2 V) is thus developed across R_4. Consequently, since the R_4 voltage is determined by the Q_2 emitter current, and the Q_2 emitter and collector currents are virtually identical, it can be seen that Q_2 acts as a constant-current generator under this condition, and that the IR LED drive currents are virtually independent of variations in supply voltage. The peak LED drive current thus approximately equals $1.2/R_4$, and R_4 (in ohms) = $1.2/I$, where I is the peak LED current in amps.

Figure 8.10 shows a 20 kHz square-wave generator (made from a 555 or 7555 'timer' IC) that can be added to the *Figure 8.9* output stage, to make a continuous-tone transmitter. In this case R_4 should be given a value of 8.2 Ω or greater, to limit peak LED currents to less than 150 mA.

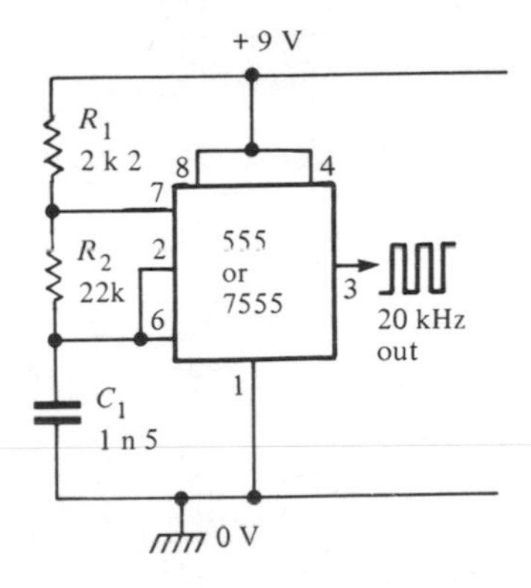

Figure 8.10 20 kHz square-wave generator

Alternatively, *Figure 8.11* shows the circuit of a tone-burst generator (giving 1 mS bursts of 20 kHz at 50 mS intervals) that can be added to *Figure 8.9* to make a tone-burst transmitter. Here, two sections of a 4011B (or CD4011) CMOS (COS/MOS) quad 2-input NAND gate IC are wired as a non-symmetrical astable multivibrator producing 1 mS and 49 mS periods; this waveform is buffered by a third 4011B stage and used to gate a 20 kHz 555/7555 astable via D2, and the output of the 555/7555 astable is then inverted via a fourth 4011B stage, ready for feeding to the transmitter output stage.

Note when using the *Figure 8.11* circuit that R_4 in *Figure 8.9* can be given a value as low as 2.2 Ω, to give peak output currents of about 550 mA, but that under this condition the transmitter will consume a mean current of little more than 6 mA; this current can be provided by either a battery or a mains-derived supply; a suitable mains-powered supply is shown in *Figure 8.12* (note that BR_1 is a bridge rectifier).

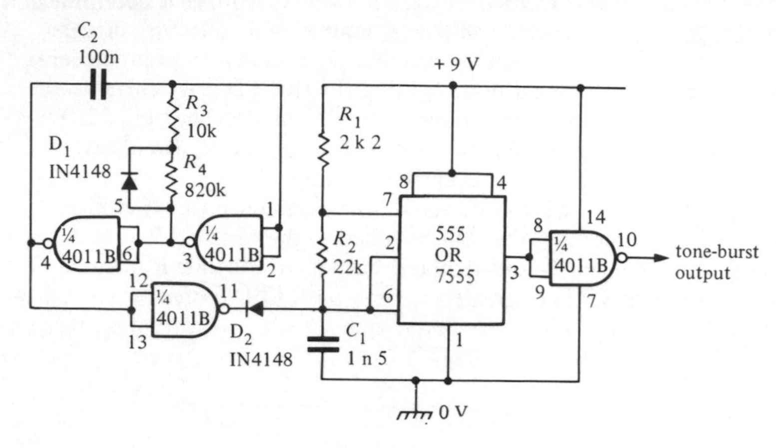

Figure 8.11 Tone-burst (1 mS burst of 20 kHz at 50 mS intervals) waveform generator

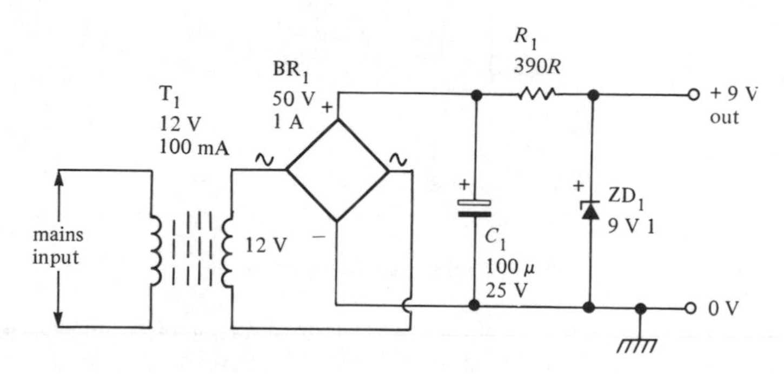

Figure 8.12 Mains-powered 9 V supply

A receiver pre-amp

Figure 8.13 shows the practical circuit of a high-gain 20 kHz 'tone' pre-amplifier designed for use in an infra-red receiver. Here, the two IR detectors are connected in parallel and wired in series with R_1, so that the detected IR signal is developed across this resistor. This signal is amplified by cascaded op-amps IC_1 and IC_2, which can provide a maximum signal gain of about ×17 680 (= ×83 via IC_1 and ×213 via

IC_2), but have gain fully variable via RV_1. These two amplifier stages have their responses centred on 20 kHz, with 3rd-order low-frequency roll-off provided via $C_4-C_5-C_6$, and 3rd-order high-frequency roll-off provided by C_8 and the internal capacitors of the ICs.

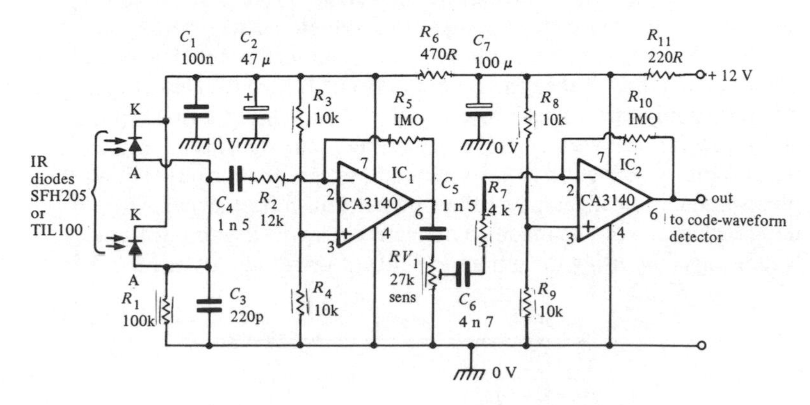

Figure 8.13 Infra-red receiver pre-amplifier circuit

The above circuit can be used with either SFH205 or TIL100 IR diodes; these devices are housed in black infra-red transmissive mouldings which reduce ambient white-light interference; *Figure 8.14* shows their outlines and pin connections.

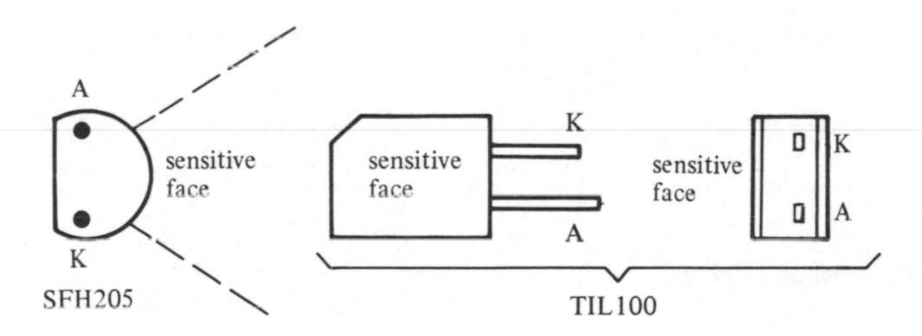

Figure 8.14 Outline and connections of the SFH205 and TIL100 IR photodiodes

The output of the *Figure 8.13* pre-amplifier can be taken from IC_2 and fed directly to a suitable code-waveform detector circuit, such as that shown in *Figure 8.15*. Note, however, that if the T_x–R_x system is to be used over ranges less than 2 metres or so the pre-amp output can be taken directly from IC_1, and all the RV_1 and IC_2 circuitry can be omitted from the pre-amp design.

Code waveform detector

In the *Figure 8.15* code waveform detector circuit the 20 kHz 'tone' waveforms (from the pre-amp output) are converted into d.c. via the $D_1-D_2-C_2-R_5-C_3$ network and fed (via R_5) to the non-inverting input of the op-amp voltage comparator, which has its inverting input connected to a thermally stable 1V0 d.c. reference point. The overall circuit action is such that the op-amp output is high (at almost full positive supply rail voltage) when a 20 kHz tone input signal is present, and is low (at near-zero volts) when a tone input signal is absent; if the input signal is derived from a tone-burst system, the output follows the pulse-modulation envelope of the original transmitter signal. The detector output can be made to activate a relay in the absence of a 'beam' signal by using the expander/output driver circuit of *Figure 8.16*.

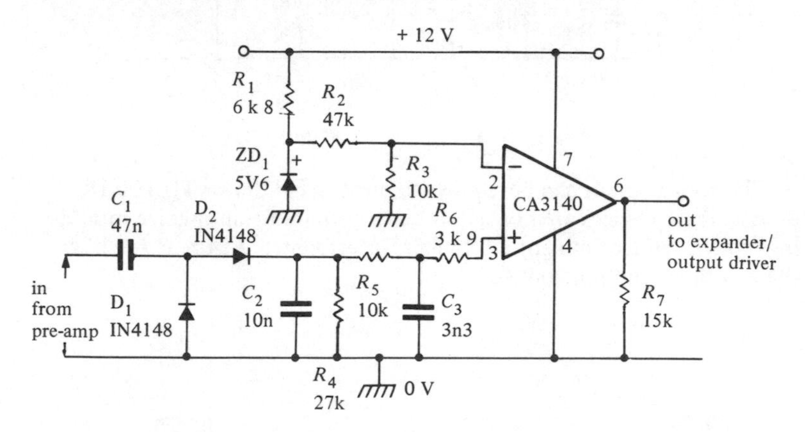

Figure 8.15 Code waveform detector circuit

Expander/output driver

The operating theory of the *Figure 8.16* circuit is fairly simple. When the input signal from the detector circuit switches high C_1 charges rapidly via D_1, but when the input switches low C_1 discharges slowly via R_1 and RV_1; C_1 thus provides a d.c. output voltage that is a 'time-expanded' version (with expansion presettable via RV_1) of the d.c. input voltage. This d.c. output voltage is buffered and inverted via IC_{1a} and used to activate relay RLA via Q_1 and an AND gate made from IC_{1b} and IC_{1c} (IC_1 is a 4011B or CD4011 CMOS quad 2-input NAND gate IC).

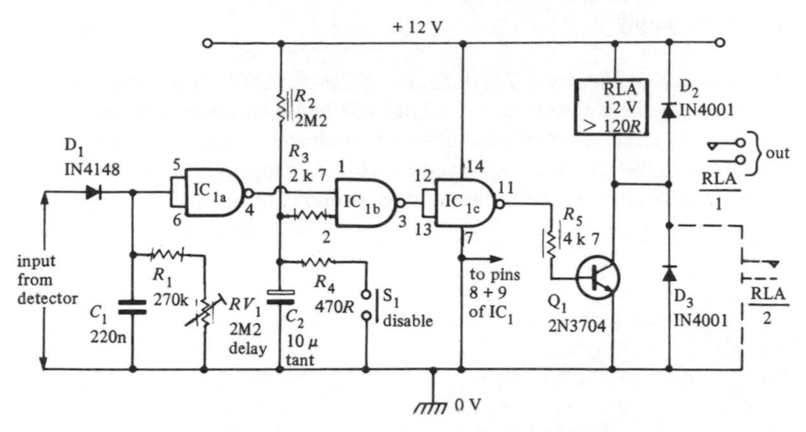

Figure 8.16 Expander/output driver circuit

Normally, the other (pin-2) input of the AND gate is biased high via R_2, and the circuit action is such that (when used in a complete IR light-beam system) the relay is off when the beam is present, but is driven on when the beam is absent for more than 100 mS or so. This action does not occur, however, when pin-2 of the AND gate is pulled low; under this condition the relay is effectively disabled.

The purpose of the $R_2 - C_2$ network is to automatically disable the relay network via the AND gate (in the way described above) for several seconds after power is initially connected to the circuit or after DISABLE switch S_1 is briefly operated, so that the owner can safely pass through the beam without activating the relay. Note that the relay can be made self-latching, if required, by wiring normally-open relay contacts RLA/2 between Q_1 emitter and collector, as shown dotted in the diagram.

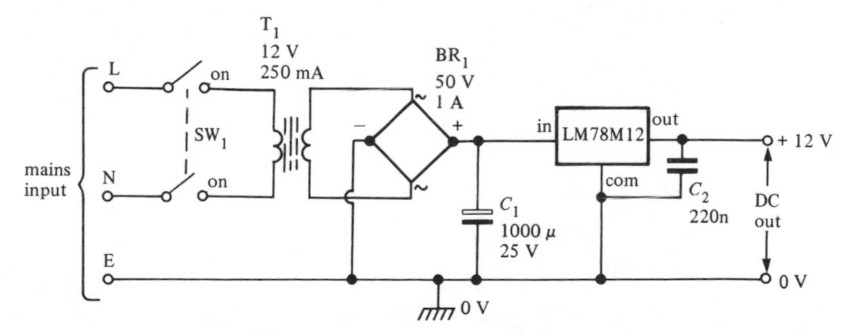

Figure 8.17 Mains-powered regulated 12 V supply

A power supply

The circuits of *Figures 8.13, 8.15* and *8.16* can be directly interconnected to make a complete infra-red light-beam receiver that can respond to either continuous-tone or tone-burst signals. Such a receiver should be powered via a regulated 12 volt d.c. supply; *Figure 8.17* shows the circuit of a suitable mains-powered unit.

APPENDIX

SEMICONDUCTOR OUTLINES AND PIN DESIGNATIONS

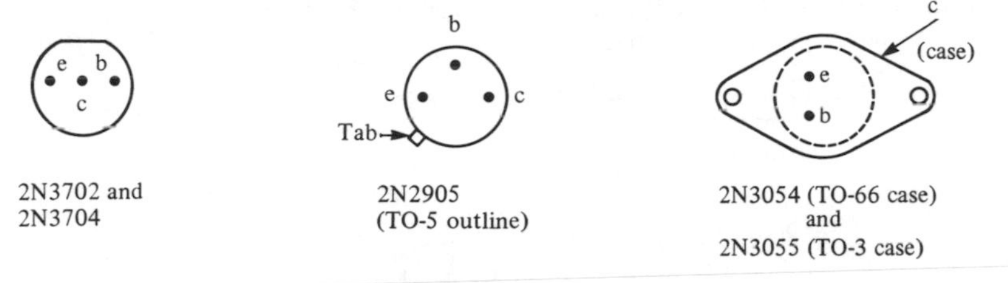

Figure A.1 Transistor details

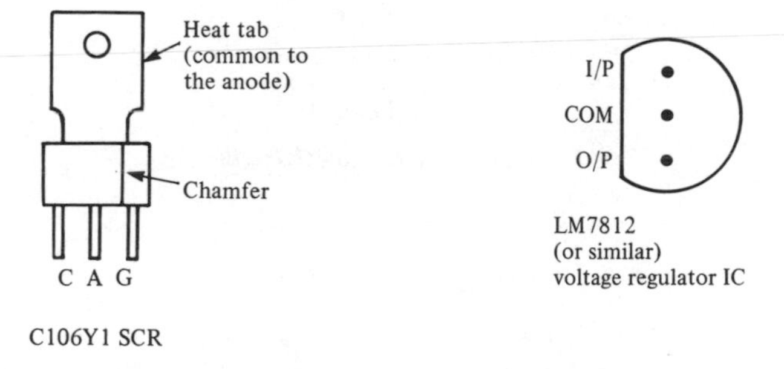

Figure A.2 SCR and voltage regulator IC details

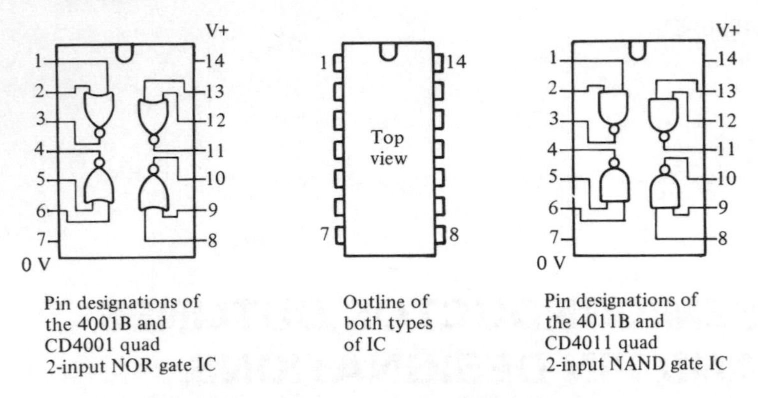

Figure A.3 CMOS (COS/MOS) IC details

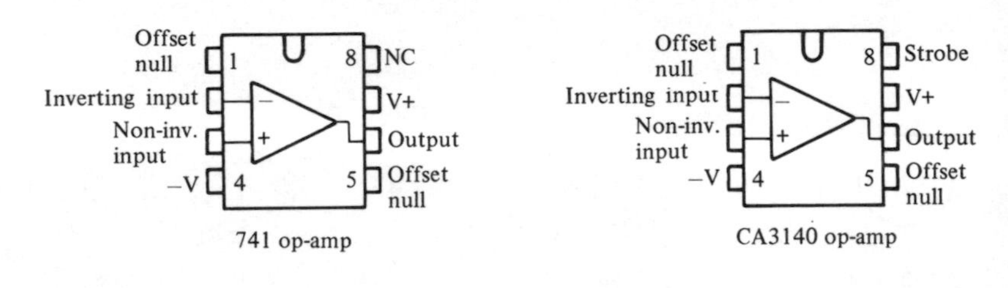

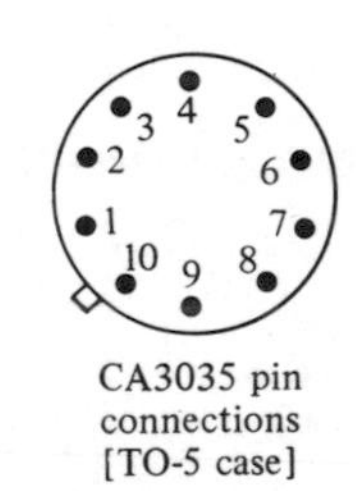

Figure A.4 Linear IC details

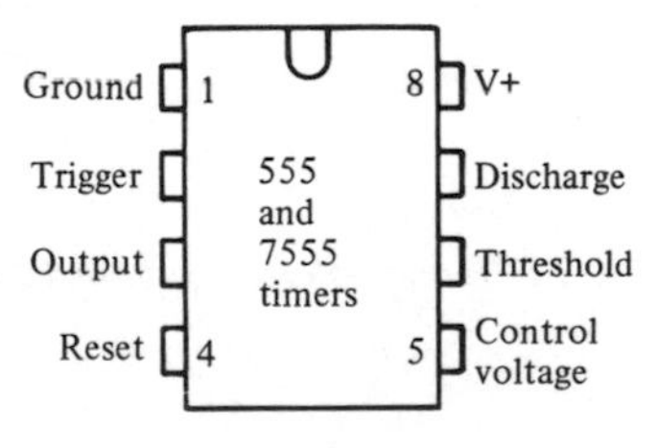

Figure A.5 Timer IC details

INDEX